INSIGHTS
IN
BIOLOGY

W9-CPE-622

TRAITS

AND

FATES

Student Manual

DEVELOPED BY
EDUCATION DEVELOPMENT CENTER, INC.

KENDALL/HUNT PUBLISHING COMPANY
4050 Westmark Drive P.O. Box 1840 Dubuque, Iowa 52004-1840

This book was prepared with the support of National Science Foundation (NSF) Grant ESI-9255722. However, any opinions, findings, conclusions and/or recommendations herein are those of the author and do not necessarily reflect the view of NSF.

Library of Congress Catalog Card Number: 96-80278

ISBN 0-7872-2209-7

Printed in the United States of America

10   9   8   7   6   5

EDC   Education Development Center, Inc.

CENTER FOR SCIENCE EDUCATION

Dear Students:

Welcome to *Insights in Biology*. This module, *Traits and Fates*, explores the relationship between what an organism actually inherits from its parents and the other biological or environmental factors that will also influence and determine the characteristics of that organism. Throughout the module, the questions "What did I inherit? What will I pass on? What will the ever-expanding scientific breakthroughs in genetics mean for me and for society?" guide the introduction and development of the concepts of inheritance. You will find that the inheritance of traits is more than the simple passing on of DNA (genes) from one generation to another, but also how those genes are expressed in the offspring. The variety you see among your classmates and in the natural world at large is due to many factors, genetic and environmental. This subdiscipline of biology is a fascinating one to explore, but it also raises many ethical and moral questions. The answers must not be left to the scientist; educated citizens must be involved in deciding the uses of research that directly involves their lives.

Glance through the pages of this manual. Your first instinct is correct: This is not a traditional biology textbook. Although textbooks provide a good deal of useful information, they are not the only way to discover science. In this Student Manual, you will find that chapters have been replaced by Learning Experiences that include readings and activities. The activities include laboratory experimentation, concept mapping, model building, simulation exercises, and a research paper. These learning experiences emphasize the processes of science and the connections among concepts in genetics, patterns of inheritance, and evolutionary theory.

One of our main goals is to engage you in the excitement of biology. The study of biology is much more than facts. It is a discipline that is as alive as the subject it portrays: new questions arise, new theories based on evidence are proposed, and new understandings are achieved. As a result of these new insights, technologies are developed which will impact your everyday lives and the kinds of decisions you will need to make. We hope that this curriculum encourages you to ask questions, to develop greater problem-solving and thinking skills, and to recognize the importance of science in your life.

Insights in Biology Staff

55 CHAPEL STREET
NEWTON, MASSACHUSETTS 02158-1060
TELEPHONE: 617-969-7100
FAX: 617-630-8439

Table of Contents

Learning Experiences

1   Finding the "Gene" in Genetics   1

Genes in the News—Activity
The Blue People of Troublesome Creek—Reading

2   No Matter What Your Shape   13

A Pea by Any Other Name Is Still a Seed—Activity
Am I a Carrier, and What does That Mean?—Case Study

3   The Language of Heredity   25

The Molecule of Life—Reading
From DNA to RNA to Protein—Activity
The DNA Shuffle—Activity
The Wonderful Mistake—Reading

4   Change in Sequence, Change in Trait   41

Simple Mistake, Serious Consequences—Activity
The Error of Our DNA—Activity
Living With Gaucher's Disease—Case Study

**5   A Gene for Everything and
     Every Gene in Its Place**   49

DNA, DNA, Wherefore Art Thou, DNA?—Activity
Characterizing Chromosomes—Activity
Reading the Future? Analyzing Chromosomes and Explaining
     Consequences—Activity

6   Legacy of Heredity   61

The Cells of Genetic Continuity—Reading
Dance of the Chromosomes—Activity
Simulating the Stages of Meiosis—Activity
The Mating Game—Activity
Mistakes of Meiosis—Activity
Cloning of the Lamb: Silence of the Man?—Reading

7 **IN THE ABBEY GARDEN**   **81**

The Flowers that Bloom in the Spring—Activity
The Parson and His Peas—Reading
My Parents Are Round and I'm Wrinkled; What Gives?—Reading
Sorting the Crosses—Reading
A Growing Concern—Case Study

8 **MAPPING GENETIC TRAILS**   **95**

I'll Trade You—Activity
Hairy, the Blue Tomato—Activity
Home, Home on the Chromosome—Reading

9 **WHAT MENDEL NEVER KNEW**   **105**

Variation: It's Not That Simple—Activity
Playing Your Hand—Reading

10 **GENETICS AND EVOLUTION: THE NEEDLE AND THE THREAD**   **115**

Onward and Upward?—Activity
Charles Darwin's Theory—Then and Now—Reading
Teddy Graham Selection—Activity
Population Genetics—Reading
What's in Our Gene Pool?—Activity
Natural Detours—Reading

11 **A CURRENT AFFAIR**   **131**

Consider the Source—Activity

APPENDIX

A   **GLOSSARY OF TERMS**   **137**

FINDING THE "GENE" IN GENETICS

PROLOGUE **W**ho are you? What determines what you look like? Can you predict how tall you will be, how healthy—or even wealthy— you may become? How much of "you" is determined by your biological inheritance, how much is influenced by the environment, and how much can you shape by your own actions and choices? As you explore the concepts in this module, your responses to these questions—and other questions of your own—may change. The activities, investigations, and readings in this module will deepen and expand your understanding of your own characteristics and what these characteristics (traits) may mean for you now and in the future.

GENES IN THE NEWS

INTRODUCTION The term "gene" is often seen in newspapers and magazines, and heard on television. You may have used it yourself in conversation or in the classroom. As is true of many words from science, the word gene has several levels of meaning, from the casual to the biochemical. In this activity, you will begin to determine what you and classmates mean by the word and compare it with the ways it is used by others, such as science writers and scientists.

▶ TASK

1. Examine the headlines pictured in Figure 1.1. Using these headlines as a guide, respond to the question "What is a gene?" by writing a definition in your notebook. Be prepared to share your definition with the class.

2. Write the class definition of "gene" in your notebook.

3. Your teacher will assign to your group one headline from Figure 1.1. Discuss with your group what you think the article is about. Then write your group's version of an opening paragraph for the article. Your paragraph should reflect the information in the headline as well as the class definition of a gene. You may be asked to read your headline and paragraph aloud.

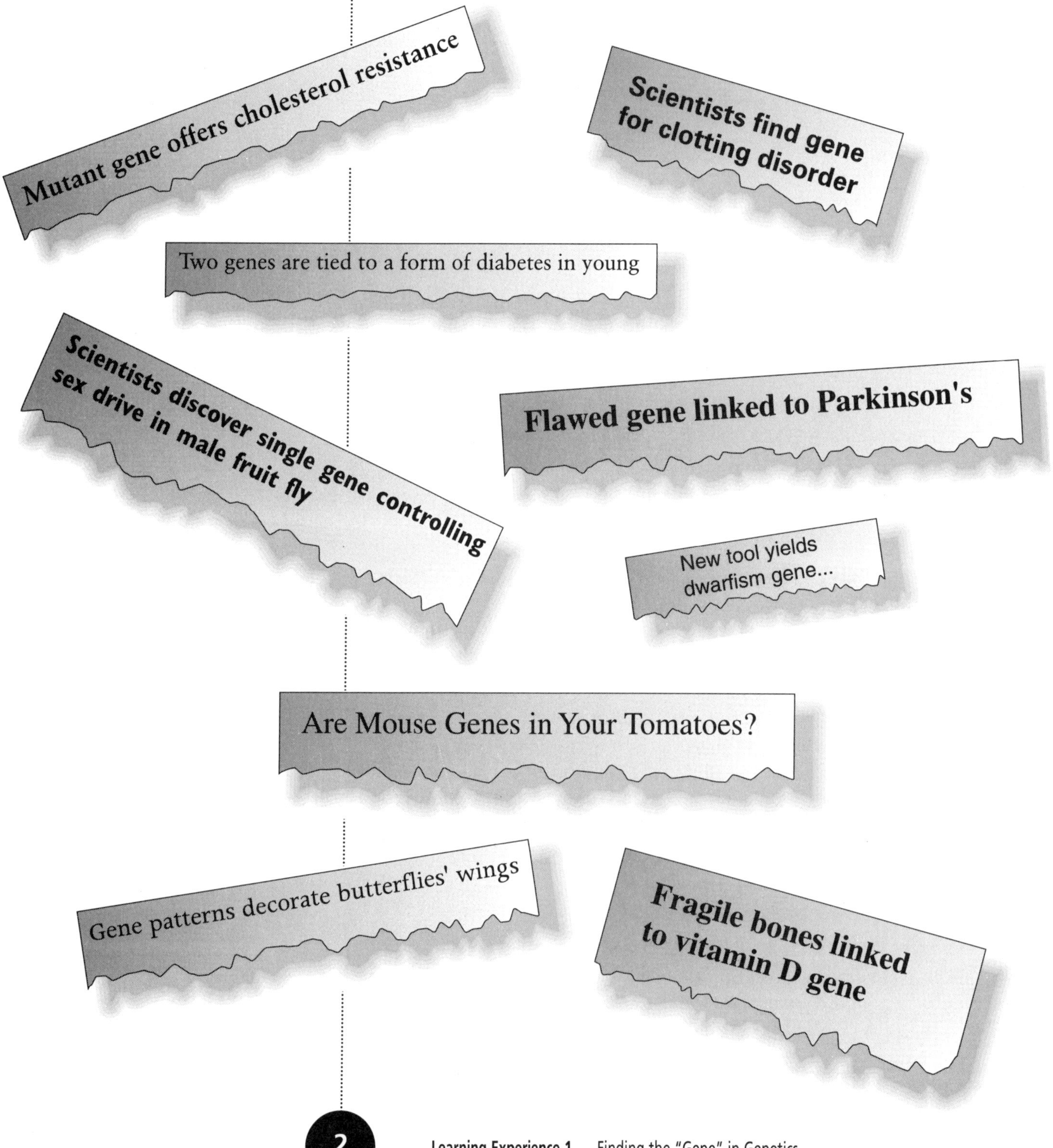

Figure 1.1
Headlines using the term "gene."

Mutant gene offers cholesterol resistance

Scientists find gene for clotting disorder

Two genes are tied to a form of diabetes in young

Scientists discover single gene controlling sex drive in male fruit fly

Flawed gene linked to Parkinson's

New tool yields dwarfism gene...

Are Mouse Genes in Your Tomatoes?

Gene patterns decorate butterflies' wings

Fragile bones linked to vitamin D gene

THE BLUE PEOPLE OF TROUBLESOME CREEK

THE STORY OF AN APPALACHIAN MALADY, AN INQUISITIVE DOCTOR, AND A PARADOXICAL CURE.

by Cathy Trost, Science 82, *November 1982, pp. 35–39.*

Six generations after a French orphan named Martin Fugate settled on the banks of eastern Kentucky's Troublesome Creek with his red-headed American bride, his great-great-great-great grandson was born in a modern hospital not far from where the creek still runs.

The boy inherited his father's lankiness and his mother's slightly nasal way of speaking.

What he got from Martin Fugate was dark blue skin. "It was almost purple," his father recalls.

Doctors were so astonished by the color of Benjy Stacy's skin that they raced him by ambulance from the maternity ward in the hospital near Hazard to a medical clinic in Lexington. Two days of tests produced no explanation for skin the color of a bruised plum.

A transfusion was being prepared when Benjy's grandmother spoke up. "Have you ever heard of the blue Fugates of Troublesome Creek?" she asked the doctors.

"My grandmother Luna on my dad's side was a blue Fugate. It was real bad in her," Alva Stacy, the boy's father, explained. "The doctors finally came to the conclusion that Benjy's color was due to blood inherited from generations back."

Benjy lost his blue tint within a few weeks, and now he is about as normal looking a seven-year-old boy as you could hope to find. His lips and fingernails still turn a shade of purple-blue when he gets cold or angry—a quirk that so intrigued medical students after Benjy's birth that they would crowd around the baby and try to make him cry. "Benjy was a pretty big item in the hospital," his mother says with a grin.

Dark blue lips and fingernails are the only traces of Martin Fugate's legacy left in the boy; that, and the recessive gene that has shaded many of the Fugates and their kin blue for the past 162 years.

They're known simply as the "blue people" in the hills and the hollows around Troublesome and Ball Creeks. Most lived to their 80s and 90s without serious illness associated with the skin discoloration. For some, though, there was a pain not seen in lab tests. That was the pain of being blue in a world that is mostly shades of white to black.

There was always speculation in the hollows about what made the blue people blue—heart disease, a lung disorder, the possibility proposed by one old-timer that "their blood is just a little closer to their skin." But no one knew for sure, and doctors rarely paid visits to the remote creekside settlements where most of the "blue Fugates" lived until well into the 1950s. By the time a young hematologist from the University of Kentucky came down to Troublesome Creek in the 1960s to cure the blue people, Martin Fugate's descendants had multiplied their recessive genes all over the Cumberland Plateau.

Madison Cawein began hearing rumors about the blue people when he went to work at the University of Kentucky's Lexington medical clinic in 1960. "I'm a hematologist, so something like that perks up my ears," Cawein says, sipping on whiskey sours and letting his mind slip back to the summer he spent "tromping around the hills looking for blue people."

Cawein is no stranger to eccentricities of the body. He helped isolate an antidote for cholera, and he did some of the early work on L-dopa, the drug for Parkinson's disease. But his first love, which he developed as an Army medical technician in World War II, was hematology. "Blood cells always looked so beautiful to me," he says.

Cawein would drive back and forth between Lexington and Hazard—an eight-hour ordeal before the tollway was built—and scour the hills looking for the blue people he'd heard rumors about. The American Heart Association had a clinic in Hazard, and it was there that Cawein met "a great big nurse" who offered to help.

Her name was Ruth Pendergrass, and she had been trying to stir up medical interest in the blue people ever since a dark blue woman walked into the county health department one bitterly cold afternoon and asked for a blood test.

"She had been out in the cold and she was just *blue!*" recalls Pendergrass, who is now 69 and retired from nursing. "Her face and her fingernails were almost indigo blue. It like to scared me to death. She looked like she was having a heart attack. I just knew that patient was going to die right there in the health department, but she wasn't a'tall alarmed. She told me that her family was the blue Combses who lived up on Ball Creek. She was a sister to one of the Fugate women."

About this same time, another of the blue Combses, named Luke, had taken his sick wife up to the clinic at Lexington. One look at Luke was enough to "get those doctors down here in a hurry," says Pendergrass, who joined Cawein to look for more blue people.

Trudging up and down the hollows, fending off "the two mean dogs that everyone had in their front yard," the doctor and the nurse would spot someone at the top of a hill who looked blue and take off in wild pursuit. By the time they'd get to the top, the person would be gone. Finally, one day when the frustrated doctor was milling inside the Hazard clinic, Patrick and Rachel Ritchie walked in.

"They were bluer'n hell," Cawein says. "Well, as you can imagine,

I really examined them. After concluding that there was no evidence of heart disease, I said, 'Aha!' I started asking them questions: 'Do you have any relatives who are blue?' Then I sat down and we began to chart the family.

Cawein remembers the pain that showed on the Ritchie brother's and sister's faces. "They were really embarrassed about being blue," he said. "Patrick was all hunched down in the hall. Rachel was leaning against the wall. They wouldn't come into the waiting room. You could tell how much it bothered them to be blue."

After ruling out heart and lung diseases, the doctor suspected methemoglobinemia, a rare hereditary blood disorder that results from excess levels of methemoglobin in the blood. Methemoglobin, which is blue, is a nonfunctional form of the red hemoglobin that carries oxygen. It is the color of oxygen-depleted blood seen in the blue veins just below the skin.

If the blue people did have methemoglobinemia, the next step was to find out the cause. It can be brought on by several things: abnormal hemoglobin formation, an enzyme deficiency, and taking too much of certain drugs, including vitamin K, which is essential for blood clotting and is abundant in pork liver and vegetable oil.

Cawein drew "lots of blood" from the Ritchies and hurried back to his lab. He tested first for abnormal hemoglobin, but the results were negative.

Stumped, the doctor turned to the medical literature for a clue. He found references to methemoglobinemia dating to the turn of the century, but it wasn't until he came across E. M. Scott's 1960 report in the *Journal of Clinical Investigation* that the answer began to emerge.

Scott was a Public Health Service doctor at the Arctic Health Research Center in Anchorage who had discovered hereditary methemoglobinemia among Alaskan Eskimos and Indians. It was caused, Scott speculated, by an absence of the enzyme diaphorase from their red blood cells. In normal people hemoglobin is converted to methemoglobin at a very slow rate. If this conversion continued, all the body's hemoglobin would eventually be rendered useless. Normally, diaphorase converts methemoglobin back to hemoglobin. Scott also concluded that the condition was inherited as a simple recessive trait. In other words, to get the disorder, a person would have to inherit two genes for it, one from each parent. Somebody with only one gene would not have the condition but could pass the gene to a child.

Scott's Alaskans seemed to match Cawein's blue people. If the condition were inherited as a recessive trait, it would appear most often in an inbred line.

Cawein needed fresh blood to do an enzyme assay. He had to drive eight hours back to Hazard to search out the Ritchies, who lived in a tapped-out mining town called Hardburly. They took the doctor to see

see their uncle, who was blue, too. While in the hills, Cawein drove over to see Zach (Big Man) Fugate, the 76-year-old patriarch of the clan on Troublesome Creek. His car gave out on the dirt road to Zach's house, and the doctor had to borrow a Jeep from a filling station.

Zach took the doctor even farther up Copperhead Hollow to see his Aunt Bessie Fugate, who was blue. Bessie had an iron pot of clothes boiling in her front yard, but she graciously allowed the doctor to draw some of her blood.

"So I brought back the new blood and set up my enzyme assay," Cawein continued. "And by God, they didn't have the enzyme diaphorase. I looked at other enzymes and nothing was wrong with them. So I knew we had the defect defined."

Just like the Alaskans, their blood had accumulated so much of the blue molecule that it overwhelmed the red of normal hemoglobin that shows through as pink in the skin of most Caucasians.

Once he had the enzyme deficiency isolated, methylene blue sprang to Cawein's mind as the "perfectly obvious" antidote. Some of the blue people thought the doctor was slightly addled for suggesting that a blue dye could turn them pink. But Cawein knew from earlier studies that the body has an alternative method of converting methemo-globin back to normal. Activating it required adding to the blood a sub-stance that acts as an "electron donor." Many substances do this, but Cawein chose methylene blue because it had been used successfully and safely in other cases and because it acts quickly.

Cawein packed his black bag and rounded up Nurse Pendergrass for the big event. They went over to Patrick and Rachel Ritchie's house and injected each of them with 100 milligrams of methylene blue.

"Within a few minutes, the blue color was gone from their skin," the doctor said. "For the first time in their lives, they were pink. They were delighted."

"They changed colors!" remembered Pendergrass. "It was really something exciting to see."

The doctor gave each blue family a supply of methylene blue tablets to take as a daily pill. The drug's effects are temporary, as meth-ylene blue is normally excreted in the urine. One day, one of the older mountain men cornered the doctor. "I can see that old blue running out of my skin," he confided.

Before Cawein ended his study of the blue people, he returned to the mountains to patch together the long and twisted journey of Martin Fugate's recessive gene. From a history of Perry County and some Fugate family Bibles listing ancestors, Cawein has constructed a fairly complete story.

Martin Fugate was a French orphan who emigrated to Kentucky in 1820 to claim a land grant on the wilderness banks of Troublesome Creek. No mention of his skin color is made in the early histories of the area, but family lore has it that Martin himself was blue.

The odds against it were incalculable, but Martin Fugate managed to find and marry a woman who carried the same recessive gene. Elizabeth Smith, apparently, was as pale-skinned as the mountain laurel that blooms every spring around the creek hollows.

Martin and Elizabeth set up housekeeping on the banks of Troublesome and began a family. Of their seven children, four were reported to be blue.

The clan kept multiplying. Fugates married other Fugates. Sometimes they married first cousins. And they married the people who lived closest to them, the Combses, Smiths, Ritchies, and Stacys. All lived in isolation from the world, bunched in log cabins up and down the hollows, and so it was only natural that a boy married the girl next door, even if she had the same last name.

"When they settled this country back then, there was no roads. It was hard to get out, so they intermarried," says Dennis Stacy, a 51-year-old coal miner and amateur genealogist who has filled a loose-leaf notebook with the laboriously traced blood lines of several local families.

Stacy counts Fugate blood in his own veins. "If you'll notice," he observes, tracing lines on his family's chart, which lists his mother's and his father's great grandfather as Henley Fugate, "I'm kin to myself."

The railroad didn't come through eastern Kentucky until the coal mines were developed around 1912, and it took another 30 or 40 years to lay down roads along the local creeks.

Martin and Elizabeth Fugate's blue children multiplied in this natural isolation tank. The marriage of one of their blue boys, Zachariah, to his mother's sister triggered the line of succession that would result in the birth, more than 100 years later, of Benjy Stacy.

When Benjy was born with purple skin, his relatives told the perplexed doctors about his great grandmother Luna Fugate. One relative describes her as "blue all over," and another calls Luna "the bluest woman I ever saw."

Luna's father, Levy Fugate, was one of Zachariah Fugate's sons. Levy married a Ritchie girl and bought 200 acres of rolling land along Ball Creek. The couple had eight children, including Luna.

A fellow by the name of John E. Stacy spotted Luna at Sunday services of the Old Regular Baptist Church back before the century turned. Stacy courted her, married her, and moved over from Troublesome Creek to make a living in timber on her daddy's land.

Luna has been dead nearly 20 years now, but her widower survives. John Stacy still lives on Lick Branch of Ball Creek. His two-room log cabin sits in the middle of Laurel Fork Hollow. Luna is buried at the top of the hollow. Stacy's son has built a modern house next door, but the old logger won't hear of leaving the cabin he built with timber he personally cut and hewed for Luna and their 13 children.

Stacy recalls that his father-in-law, Levy Fugate, was "part of the family that showed blue. All them old fellers way back then was blue. One of 'em—I remember seein' him when I was just a boy—Blue Anze, they called him. Most of them old people went by that name—the blue Fugates. It run in the generation who lived up and down Ball [Creek]."

"They looked like anybody else 'cept they had the blue color," Stacy says, sitting in a chair in his plaid flannel shirt and suspenders, next to a cardboard box where a small black piglet, kept as a pet, is squealing for his bottle. "I couldn't tell you what caused it."

The only thing Stacy can't—or won't—remember is that his wife Luna was blue. When asked about it, he shakes his head and stares steadfastly ahead. It would be hard to doubt this gracious man except that you can't find another person who knew Luna who doesn't remember her as being blue.

"The bluest Fugates I ever saw was Luna and her kin," says Carrie Lee Kilburn, a nurse who works at the rural medical center called Homeplace Clinic. "Luna was bluish all over. Her lips were as dark as a bruise. She was as blue a woman as I ever saw."

Luna Stacy possessed the good health common to the blue people, bearing at least 13 children before she died at 84. The clinic doctors only saw her a few times in her life and never for anything serious.

As coal mining and the railroads brought progress to Kentucky, the blue Fugates started moving out of their communities and marrying other people. The strain of inherited blue began to disappear as the recessive gene spread to families where it was unlikely to be paired with a similar gene.

Benjy Stacy is one of the last of the blue Fugates. With Fugate blood on both his mother's and his father's side, the boy could have received genes for the enzyme deficiency from either direction. Because the boy was intensely blue at birth but then recovered his normal skin tones, Benjy is assumed to have inherited only one gene for the condition. Such people tend to be very blue only at birth, probably because newborns normally have smaller amounts of diaphorase. The enzyme eventually builds to normal levels in most children and to almost normal levels in those like Benjy, who carry one gene.

Hilda Stacy is fiercely protective of her son. She gets upset at all the talk of inbreeding among the Fugates. One of the supermarket tabloids once sent a reporter to find out about the blue people, and she was distressed with his preoccupation with intermarriages.

She and her husband Alva have a strong sense of family. They sing in the Stacy Family Gospel Band and have provided their children with a beautiful home and a menagerie of pets, including horses.

"Everyone around here knows about the blue Fugates," says Hilda Stacy who, at 26, looks more like a sister than a mother to her children. "It's common. It's nothing."

"It's common. It's nothing."

Cawein and his colleagues published their research on hereditary diaphorase deficiency in the *Archives of Internal Medicine* in 1964. He hasn't studied the condition for years. Even so, Cawein still gets calls for advice. One came from a blue Fugate who'd joined the Army and been sent to Panama, where his son was born bright blue. Cawein advised giving the child methylene blue and not worrying about it.

The doctor was recently approached by the producers of the television show "That's Incredible." They wanted to parade the blue people across the screen in their weekly display of human oddities. Cawein would have no part of it, and he related with glee the news that a film crew sent to Kentucky from Hollywood fled the "two mean dogs in every front yard" without any film. Cawein cheers their bad luck not out of malice but out of a deep respect for the blue people of Troublesome Creek.

"They were poor people," concurs Nurse Pendergrass, "but they were good."

▶ ANALYSIS

Write responses to the following in your notebook.

1. What physical trait did Martin Fugate and his wife pass on to their many generations of offspring?

2. Why did this condition occur?

3. What role did genes play in both the appearance of the trait and the passing on of the trait?

4. Based on this reading and your responses to the above questions, how might you begin to define the term "genetics"?

EXTENDING IDEAS

ON THE JOB

GENEALOGIST   Philip is a genealogist; that is, he puts together people's family trees. Recently he was approached by a woman who was trying to piece together some information she had about a once-hidden family secret. Louisa Kay, his client's great-aunt, had had an affair with a married man. She became pregnant, and her family suddenly moved to another state, where Louisa gave birth to a girl named Sally. The client knew some information about Sally, but very little. There were two things she wanted to find out: Who was Sally's father? If Sally had any children, where were they now?

Philip knew that parents' names were usually listed on birth and death certificates as well as marriage certificates. At that time, though, Louisa and her family may not have wanted the father to be known. Regardless, Philip needed to get his hands on copies of those certificates. He started with a couple of clues given by his client. He knew that Louisa Kay originally lived in St. Paul, Minnesota. He also knew that Sally had grown up in Topeka, Kansas, married a man named Brown, and died there.

He started in St. Paul. By checking the 1900 census records, kept at branches of the National Archives in each state, he was able to discern that Louisa was living at home with her parents at that time. She had no children then. The 1910 census records had no listing of a Kay family. Philip then traveled to Topeka, where he checked the 1910 census records of that city. Louisa was listed, along with a two-year-old daughter named Sally. Bingo! He then estimated the date of birth and went to the city hall in Topeka for Sally's birth certificate. There was no birth certificate for a Sally Kay in that city, which meant she was born somewhere else. The next place to look was at her death certificate. He traveled around Topeka speaking with cemetery administrators until he found her grave. The gravestone gave her date of death. With that, he returned to the city hall and retrieved her death certificate. The line for "father" said "unknown," but it wasn't a dead end. The death certificate listed a place of birth— Wichita.

Philip flew to Wichita, where he was able to get a copy of Sally's birth certificate. Again, no name was given for her father. It was possible that there would be no way of knowing.

Putting that question aside, Philip proceeded to look for Sally's children. Since he had already found out her date of death, he went to the Topeka Public Library and checked out the obituaries on microfiche. Listed in Sally's obituary were five children: Richard of Topeka, Robert of San Francisco, Nancy of Topeka, Louisa of Wichita, and Mary of Topeka.

He'd solved one puzzle. And a single call to one of her children solved the other—they knew the name of their grandfather.

Philip enjoys his job, which to him, is like a mystery whose solution is constantly evolving. He likes the contacts with many people of widely different backgrounds, and he especially enjoys the feeling of accomplishment after solving a puzzle. Some clients are interested in the in-depth creation of a family tree. This could mean going back many generations, finding names of relatives, their children, dates of major life events, etc. Others are more interested in a specific piece of information such as the location of an heir, the date of someone's U.S. citizenship, or the place where ancestors first set foot on U.S. soil. Finding this information, as Philip is well aware, involves a lot of investigation and research. Experience is very important when becoming a professional genealogist. Philip learned much of what he knows from reading books, attending lectures by other professional genealogists, and other workshops and asking a lot of questions while visiting the different branches of the National Archives. He prefers to work independently, although there are companies worldwide which specialize in genealogy.

NO MATTER WHAT YOUR SHAPE

PROLOGUE The easiest traits to describe in an organism are the visible ones. For instance, when distinguishing between two people, you may use the the texture of their hair as an easily identifiable trait; one may have curly hair, and the other straight hair. In using differences in hair texture you have described *variants* of a single trait, that is, hair texture.

What is responsible for variants in traits? Are traits and their variants only characteristics that are directly observable, or are there underlying causes for these variants? In this learning experience, you will use variations in the shape of peas as a simple model for investigating the answers to these questions. You will then explore a trait in humans, sickle cell trait, as another example of variations in traits.

A PEA BY ANY OTHER NAME IS STILL A SEED

INTRODUCTION How might you begin to answer the question "Why is one pea wrinkled and another round?" Think about ways you could find out more about the differences between round and wrinkled peas. Are there differences in how the peas are constructed? How might these differences produce a wrinkled shape instead of a round shape? Are there differences in what the seed is made of? How would this affect the shape? In this learning experience, you will be investigating the causes of variations in traits of organisms, using these pea variants as a model.

Figure 2.1
Structure of a seed.

Before you begin to investigate the differences between wrinkled peas and round peas (which are actually seeds), you need to understand the general structure of a seed (shown in Figure 2.1). A seed is the part of a plant that results from the fertilization of the female egg by the male pollen. Following fertilization, the embryonic plant develops within a protective seed coat. In addition to the embryonic plant and the seed coat, the seed also contains a source of food that the germinating seedling will use until it is able to carry out photosynthesis. In one type of plant, this food source is in a separate structure within the seed called the endosperm. In another type of plant, the protein, starch, and fats are stored in two large seed leaves. Regardless of the system of storage, the newly sprouted plant is dependent on these stored food sources until it can make its own.

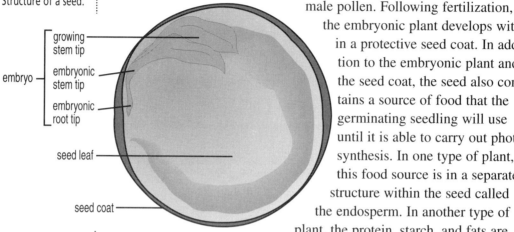

Seeds have a very low water content. During the final stages of development of the seed, cells within the seed dehydrate; that is, most of the water in the seed is removed. This results in low water content in the seed which causes most cellular processes to slow down or stop. In this dehydrated state, the embryonic plant can remain dormant but viable within the seed for long periods of time without growing or developing. This strategy enables seeds to delay germination until environmental conditions are suitable for its growth. When conditions become favorable water enters the seed, rehydration triggers the reactivation of normal metabolic processes, and germination begins.

As the plant begins to develop, it utilizes the starch stored in the endosperm or seed leaves as an energy source. It can use photosynthesis to provide its required food and energy only after breaking free of the soil and into sunlight (see Figure 2.2). In this investigation you will determine why some seeds (peas) are wrinkled at the end of their development and some are round.

Figure 2.2
(a) During seed development, water is lost from cells. This dehydration slows down metabolic processes. (b) During germination, water enters the seed, metabolic processes are reactivated, and the embryonic plant begins to grow.

a) seed development

water removed

b) germination

water added

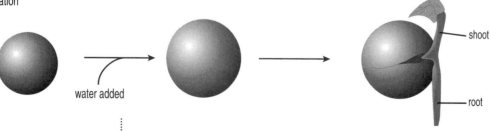

▶ MATERIALS NEEDED

For each pair of students:

- 2 pairs of safety goggles
- 10 round peas
- 10 wrinkled peas
- 1 balance
- 2 small beakers or containers (50-mL)
- 1 wax marking pencil
- 2 microscope slides with coverslips
- access to a compound microscope
- 1 razor blade or scalpel
- 1 forceps
- 1 dropping bottle of dilute Lugol's iodine
- paper towels

▶ PROCEDURE

PART A:

1. **STOP & THINK** Create a hypothesis as to why some peas are wrinkled and some are round. Record the hypothesis in your notebook.

2. Read steps 2–8 and then create a data chart in your notebook. Fill it in as you carry out this part of the experiment.

3. Weigh all 10 round peas together and record the weight of the peas on your data chart. Weigh all 10 wrinkled peas together and record the weight of the peas on your data chart.

4. Use a wax marking pencil to label two beakers "R" and "W." Also mark the beakers with the name of someone in your group. Place the dried round peas in the "R" beaker and the dried wrinkled peas in the "W" beaker, and add water until the beakers are 3/4 full.

5. Place the beakers in the location designated by your teacher until the next class session.

6. After soaking the peas overnight, label two paper towels "R" and "W." Pour off the excess water from each beaker and empty the peas from each beaker in a pile on the appropriate paper towel.

7. Determine the soaked weights of each of the two kinds of peas. Record the weights in your table.

8. Determine the weight difference for each kind of pea. Calculate the percentage of increase in weight for each kind of pea. Record the results in your table.

$$\frac{\text{weight difference}}{\text{dry weight}} \times 100 = \% \text{ increase}$$

9. **STOP & THINK** On the basis of this experiment and your understanding of seed formation, do you wish to change your hypothesis as to why some peas wrinkle and others remain round when they are dried? Record your response in your notebook.

PART B:

1. Use a wax marking pencil to label two microscope slides "R" and "W."

2. Place one drop of dilute iodine on each slide.

3. Hold a soaked wrinkled pea with forceps and cut the pea in half with a scalpel or razor blade. Cut a very thin segment or slice from the inside of the pea. Gently place the slice into the appropriate drop of dilute iodine (see Figure 2.3).

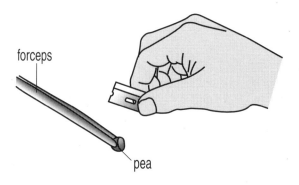

forceps

pea

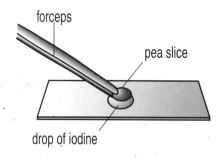

forceps

pea slice

drop of iodine

Figure 2.3
Place the cut slice of the pea into the drop of iodine on the slide.

4. Carefully wipe the blade of the scalpel or razor blade clean with a paper towel and repeat Procedure step 3 with a soaked round pea.

5. Place a cover slip at an angle over each drop and gently lower it. Observe each slice under the microscope.

6. **STOP & THINK** Describe and draw in your notebook what you see. Compare the shapes, colors, patterns, and densities of the starch grains. Look at a few slides prepared by other class members and compare them with yours.

► ANALYSIS

Write responses to the following in your notebook.

1. Write a laboratory report of your investigations. In this report you should:
    – identify the question(s) asked,
    – include the description of the experimental procedures,
    – explain what each part of the investigation was designed to do,
    – record the results, and
    – state any conclusions you can make based on your results.

2. From your conclusions decide whether you have enough data and knowledge to identify the cause of the shape difference between the two different kinds of peas and explain why or why not.

AM I A CARRIER, AND WHAT DOES THAT MEAN?

Figure 2.4
Normal red blood cells are shaped like disks. Some red blood cells of sickle cell patients become stiff and sickled. The misshapen cells often get stuck in small blood vessels, causing extreme pain and damage.

It was Health Day at Denzel Jones' high school. "Career Day, Health Day, Environment Day," fumed Denzel. "When am I going to hear about stuff that matters to me?" Denzel's class filed into the auditorium. The air was buzzing with conversation about music, friends, the last biology exam, anything except the topic of health. Who cared, anyway? Well, at least it got them out of fifth period.

As several individuals from the local health clinic talked, Denzel found himself drawn in by some topics—such as exercise, smoking, and methods for the prevention of infectious diseases. One topic particularly caught his attention because he actually knew a couple of people with the problem. A physician's assistant began to talk about something called "hemoglobinopathies," and about one in particular, sickle cell anemia. Denzel's uncle Jamal (his father's brother) had the disorder and suffered from fatigue and bouts of intense joint pain. Because it bothered him to watch his uncle suffer, Denzel was curious about the cause.

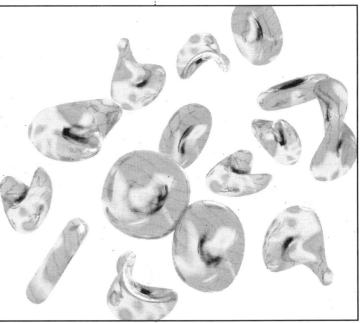

Denzel learned that *sickle cell anemia* is a disorder of red blood cells that can run in families. It causes the red blood cells to collapse into shapes resembling sickles (see Figure 2.4) when the oxygen level of the blood is low.

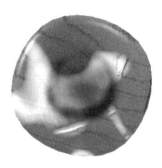

(a)

(b)

Figure 2.5

(a) Normal hemoglobin (A) remains dissolved in the cell after the release of oxygen. Cells remain disk-shaped. (b) Sickling hemoglobin (S) comes out of solution after the release of oxygen, forming long crystals and distorting the cell shape.

Red blood cells sickle because they contain hemoglobin that is biochemically a little different from the normal hemoglobin protein. Normal hemoglobin (or hemoglobin A) is found in solution in red blood cells. It binds oxygen and transports it throughout the body. Once it releases the oxygen, the hemoglobin remains in solution in the red blood cell. A variant form of hemoglobin, sickling hemoglobin, designated S, differs from normal hemoglobin only by a single amino acid. That slight difference in structure, however, alters its function. Hemoglobin S binds oxygen and carries it to where it is needed, but a problem arises when the oxygen is released and the concentration of oxygen around the hemoglobin is reduced. Though normal hemoglobin remains in solution under these conditions, the sickling hemoglobin comes out of solution; its molecules bind together into long fibrous chains (crystallizes). These fibers push out against the inside of the membrane of the red blood cell, producing the characteristic sickle shape (see Figure 2.5).

Because of their shape, these cells cannot flow easily through the tiny capillaries (the smallest passageways of the circulatory system); they get stuck and clog the flow of blood. This blockage decreases the blood supply to the vital organs—such as the heart, spleen, kidneys, and brain—and these organs can be damaged. The buildup of pressure behind the blockage also can cause small blood vessels to burst, resulting in internal bleeding and pain.

Although the symptoms of sickle cell anemia are quite variable, general features include jaundice (yellowing of the skin and other tissues due to the breakdown products of red blood cells), anemia, and pain. Infants and children may have a predisposition to infection. In later years, blood-rich organs such as the heart, spleen, and liver are damaged by the restricted blood flow; the disease may cause leg ulcers, anemia, kidney failure, stroke, and heart failure. The severity of the symptoms vary from individual to individual. Some show few symptoms, others die young.

The physician's assistant explained that sickle cell anemia is an inherited disease. The variant can run in families, and individuals can pass the variant to their children without having symptoms themselves. That is, parents who do not have sickle cell anemia can have both children with the disorder and children without the disorder. About 2.5 million, or one in every 12, African-Americans (the group most affected in the United States) carry the sickling trait without having the disease. That is, they have both kinds of hemoglobin in their red blood cells. Individuals who have both kinds of proteins are called *carriers*. Approximately 80,000 African-Americans have only sickling hemoglobin and, therefore, demonstrate the characteristics or symptoms of sickle cell anemia.

Denzel began to wonder whether anyone else in his family besides Uncle Jamal had sickle cell anemia, and whether he could be one of the individuals who had the sickle hemoglobin variant but didn't show it. The physician's assistant told the group that an easy test for sickling hemoglobin could be done at the clinic and encouraged the students to have it done.

Denzel decided he wanted to be tested.

After the assembly, Denzel approached the physician's assistant to ask questions about the test. He told Denzel that there is a test to distinguish normal hemoglobin (hemoglobin A) from sickling hemoglobin (hemoglobin S). This test is based on the understanding that the difference between the two types of hemoglobin is only one amino acid, which changes the electrical charge on the molecule. This charge difference causes the two different forms of hemoglobin to separate in an electric field: In a solution through which an electrical current is passed, hemoglobin A will move in one direction and hemoglobin S will travel the opposite way.

Denzel was amazed. The difference between being healthy and having the symptoms of sickle cell anemia was a single amino acid. And, through a fairly simple blood test, Denzel could learn whether he had any hemoglobin S.

At dinner that night Denzel recounted to his family what he had learned that day about the sickle cell trait. He said he would like to be tested and thought that it might be a good idea for everyone to be tested, to know whether they carried the trait. Denzel's father was not so sure. He worried that if he carried the trait and someone at work found out, they might think he wasn't healthy enough to operate the forklift he drove every day. And what if he applied for more health insurance? What impact would being a carrier have on that? Denzel assured him that the physician's assistant said that individuals who carried the trait rarely exhibited any symptoms of the disease and were never considered "sick." Anyway, the results of the test were confidential. No one was ever supposed to know.

Tara was worried for a different reason. She was planning to be married soon and very much wanted to have children. What if she and her fiancé, Carlos Jackson, were both carriers? What would that mean for the children they might have? She wasn't sure she wanted to know.

In the end, everyone in the family, including Denzel's four grandparents Grandpa and Grandma Jones and Grandma and Grandpa Beausejour, his sisters, and Uncle Jamal decided to be tested. Even Tara's fiancé, Carlos, wanted to find out whether he carried the trait.

Everyone nervously waited a week for the blood test results. The data in Table 2.6 on the next page were collected on the Jones and Beausejour families. (A + sign means the individual has that form of hemoglobin, and a – sign means that it was not present.)

Table 2.6
Hemoglobin Data.

Individual	Hemoglobin A	Hemoglobin S
Grandpa Jones	+	+
Grandma Jones	+	+
Grandpa Beausejour	+	–
Grandma Beausejour	+	–
Mr. Jones	+	+
Mrs. Beausejour-Jones	+	–
Uncle Jamal Jones	–	+
Tara Jones	+	+
Tabitha Jones	+	–
Denzel Jones	+	–
Carlos Jackson	+	+

▶ ANALYSIS

Write responses to the following in your notebook.

1. In this learning experience you have found out that the difference in shape of peas is the result of a difference in a single enzyme which functions during pea development. Explain the differences in the biochemistry of sickling and normal hemoglobin. How do these differences result in the visible trait (as seen under the microscope)?

2. The results of Tara's test, as well as some of the others tested, indicate that some of her red blood cells carry hemoglobin S as well as normal hemoglobin. These individuals are carriers of this variant form of protein. Yet none of them has shown any symptoms of sickle cell anemia under normal circumstances. How do you explain this?

3. Carlos enjoys mountain climbing. On occasion, at very high altitudes he has suffered fatigue and severe cramps in his joints. What do you think is the reason for this, based on his test results?

4. Scientists use family trees or *pedigrees* as a tool to record and track inherited characteristics in families. What specific characteristics have you seen in members of a family that help identify them as belonging to that family?

5. Using the test results shown in Table 2.6, create a pedigree for Denzel's family which indicates how members are related and how the sickle cell trait runs in the family.

To help you diagram the trait of sickle cell in Denzel's family, you will need to use the following symbols to create a pedigree of his family:

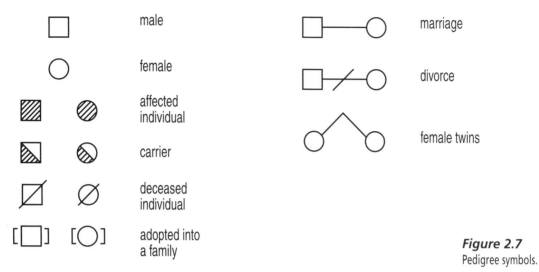

□   male

○   female

▨ ◍   affected individual

◪ ◒   carrier

▱ ⊘   deceased individual

[□] [○]   adopted into a family

□—○   marriage

□—/—○   divorce

○⌃○   female twins

Figure 2.7
Pedigree symbols.

The generations of a family are marked with Roman numerals, beginning with the first generation pictured on the chart. Each individual within a generation is labeled with an Arabic numeral (1, 2, 3, 4, etc.). Within the children of a particular couple, the first born child is usually placed to the far left, with subsequent children following to the right. Figure 2.8 is one example of a pedigree. Examine the pedigree. What can you tell about the relationships in this family? Who has the disease? Who are the carriers?

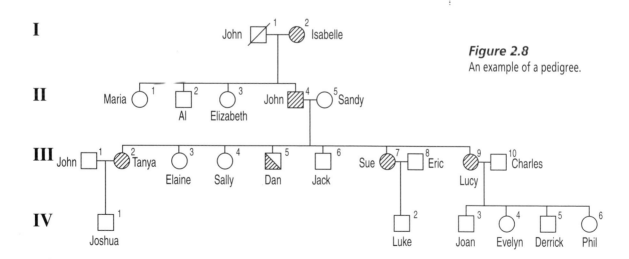

Figure 2.8
An example of a pedigree.

6. Tara and Carlos hope to marry soon and to have children. What do you think the test results mean for them? Write responses to the following:
   – List all the choices that Tara and Carlos have with respect to having children.
   – Describe all the consequences for each of the choices you just listed.
   – In a short paragraph describe what choice you might make if you were in the same situation as Tara and Carlos. Include your reasons for making that choice.
   – What do you think would happen if everyone who was confronted with this situation made the same choice you made? Write a short paragraph describing what this future might look like.
   – List four important values that influenced your decision, and explain how they influenced you. For example, some values might include religious reasons, your view of community, your sense of responsibility, your own personal health issues, and your sense of family.

EXTENDING IDEAS

In the 1970s Susan Perrine, a young doctor working in Saudi Arabia observed that many of the Arab patients who came to her clinic had surprisingly mild cases of sickle cell anemia; in fact many of them displayed no symptoms even though their blood showed the characteristic sickling effect under conditions of low oxygen. When their hemoglobin was examined the patients displayed high levels of fetal hemoglobin—the kind of hemoglobin that all humans produce before birth but generally is replaced after birth by adult hemoglobin. Fetal hemoglobin has a higher affinity for oxygen; that is, it binds oxygen more tightly than adult hemoglobin does. Apparently in these Arab patients for some reason the red blood cells had not completely switched from making fetal hemoglobin to adult hemoglobin, and the presence of this fetal hemoglobin reduced or eliminated the problem found when an individual makes only hemoglobin S. Explain why the presence of fetal hemoglobin may mask or dominate the effects of the sickling hemoglobin. Describe how this information might be used to treat sickle cell patients.

The study of genealogy, that is, tracing a family's history, can be a fascinating one. Some people track their ancestors when an unfortunate illness shows up in the immediate family and they are concerned as to whether they or their children may inherit the disease. Others search for the names and places of origin in their mother's and father's pasts for clues to their heritage. Create your own family tree. You may wish to interview your oldest relatives for their views of life and of family in past times in order to recapture family history which is often lost.

IN THE JOB

PHLEBOTOMIST   It's early in the morning, but the hospital is already busy. Metal trays covered with vials, syringes, tourniquets, and doctors' orders are being wheeled from room to room. One of these trays is followed closely by Arzu, a phlebotomist. On her morning rounds, she has orders to draw blood from an elderly woman being treated for a blood clot, a middle-aged man needing tests to find out why he has been feeling so ill, and a young girl who is in the hospital for gall bladder surgery.

Arzu loves meeting new people. Many of those she deals with aren't too happy to see her, because it is common for people to be afraid of needles. But Arzu comforts them. By educating them about what is going to happen and describing each step, she can usually quell patients' fears and take samples of their blood without any problem. Long-time patients are relieved when they see her face in the morning, knowing she cares.

Patients are thankful for Arzu's gentle touch, but they are often not aware of her great range of knowledge. She is very skilled. To complete her certificate program in phlebotomy, Arzu was trained in collecting, transporting, handling and processing blood samples; identifying and selecting equipment, supplies, and additives used in blood collection; recognizing and adhering to infection control and safety procedures; and recognizing the importance of each step from drawing blood to analysis and seeing how her part fits into the whole picture of a specific person's medical care.

Arzu's expertise in the field is in drawing blood for analysis and translating the doctors' orders for the lab technicians who do the analysis. When doctors, physician's assistants, and nurses receive the results, they use the data from these blood tests to prescribe medication and a plan for care. By being specifically trained in bloodletting procedures, Arzu's presence allows doctors and nurses the time to complete important paperwork, update records, and continue patient care toward a speedy recovery.

THE LANGUAGE OF HEREDITY

PROLOGUE In the previous learning experience, you saw that the variations in the trait of pea shape are the result of the ability of an enzyme to function: When SBEI catalyzes the bonding of smaller sucrose molecules into the larger amylopectin molecule the pea is round. If it cannot carry out this activity, the pea is wrinkled.

If a protein is responsible for the expression of a trait in an organism, what is the blueprint that directs the formation of the protein and where is this blueprint found in the cell? You may already be familiar with the molecule *deoxyribonucleic acid (DNA)* as the agent of heredity. In this learning experience you will explore in detail the structure and function of DNA and deepen your understanding of what a gene is and what it does.

THE MOLECULE OF LIFE

READING

IDENTIFYING DNA'S FUNCTION

In 1866, an Austrian monk by the name of Gregor Mendel had collected an enormous body of data on the inheritance patterns of various traits in peas. From the data, he concluded that some units he called "factors" were responsible for the passage of traits from one generation to the next. The science community started to look for the "factors," and in 1909, Wilhelm Johannsen renamed the factors *genes* (meaning "give birth to" in Greek). Soon after, William Bateson coined the term *genetics* as the study of genes.

What was the genetic substance that passed on information, that determined an offspring's visible traits, and also accounted for the incredible diversity to be found among living things? Some scientists thought it must be a protein, because proteins are present in large quantities in the cell and carry on numerous functions. Proteins are made of 20 different subunits called *amino acids* which can be joined in a great variety of combinations. Supporters of proteins as the molecule of heredity thought that this variety would allow for the diversity we see in organisms, much as the 26 letters of the English alphabet placed in a variety of ways produce an immense quantity of words.

Other scientists, noting that large amounts of DNA were also present in cells, thought that DNA was the molecule of heredity. However, it seemed too simple a molecule, with only six subunits: *deoxyribose* (a sugar), *phosphate,* and four *nitrogenous bases*. The bases *adenine* (A) and *guanine* (G) were characterized by two nitrogen rings *(purines)*; and *cytosine* (C) and *thymine* (T) had one nitrogen ring *(pyrimidines)*. The debate raged hot and heavy: Which was the molecule of heredity?

In 1928, Frederick Griffith demonstrated that a trait could be transferred from one kind of bacteria to another. He called the substance which carried this information "transforming factor," but he was unclear as to its nature. In 1943, after many years of chemical analysis and experimentation, Oswald Avery and his co-workers showed that it is DNA that directs the expression of traits within an organism and their transmission from generation to generation. But their experimental work and conclusion still did not satisfy all scientists.

Alfred Hershey and Martha Chase ended the controversy in 1952 with their experiments on bacteriophage, viruses that infect bacteria and are made of only protein and DNA. Upon infecting a bacterium, a bacteriophage produces many copies of itself (progeny) that have all the characteristics of the original infecting virus. But was it the DNA or the protein which transmitted the information for these characteristics to the progeny? To determine the answer to this question, Hershey and Chase conducted the following pivotal experiment. Knowing that the element sulfur is found only in protein and phosphorus only in DNA, they infected one dish of bacteria with a virus containing radioactive sulfur and a second dish with a virus containing radioactive phosphorus. These radioactive labels, therefore, could distinguish DNA from proteins. When the two samples of infected bacteria were assayed to see which radioactive substance was taken into the bacterial cells and used to make more viral progeny, it was found that the phosphorus entered the bacteria while the sulfur remained outside. Since only the DNA entered the cell, it was conclusively proven that DNA was the molecule used by the virus to make more of itself.

The function of DNA as the storage site and transmitter of information for traits was now known. But, what was its structure? How were those six subunits arranged in the molecule so that DNA could be the bearer of vast amounts of information and code for the incredible diversity seen in living things? Erwin Chargaff added an important piece to the puzzle with his experiments showing that the proportions of nitrogenous bases found in DNA were the same in every cell of an organism in a given species, but that the proportion varied from species to species. Examine the chart of his results in Table 3.1.

Table 3.1

COMPOSITION OF DNA IN SEVERAL SPECIES				
SOURCE	PURINES		PYRIMIDINES	
	Adenine	Guanine	Cytosine	Thymine
Human	30.4%	19.6%	19.9%	30.1%
Ox	29.0%	21.2%	21.2%	28.7%
Salmon Sperm	29.7%	20.8%	20.4%	29.1%
Wheat germ	28.1%	21.8%	22.7%	27.4%
E. coli	24.7%	26.0%	25.7%	23.6%
Sea urchin	32.8%	17.7%	17.3%	32.1%

Note the proportion of each nitrogenous base in human DNA. What do you notice about the relative (not exact) proportions of the four nitrogenous bases? Does this observation hold true for the other species as well? What conclusion might you draw about which nitrogenous base might be joined to another nitrogenous base?

DISCOVERING THE STRUCTURE OF DNA

In the early 1950s, James Watson, an American, and Francis Crick, an Englishman, set out together to discover the structure of the DNA molecule—that is, in what molecular configuration the six subunits were arranged. They did not carry out laboratory experiments themselves, but based their hypothesis on data accumulated by others, including the X-ray work of Rosalind Franklin which showed that DNA had a double helical (or spiral) form (see Figure 3.2). From this information, they were able to build a model that satisfied all the known data. They worried that the structure would be "dull," that is, that it could not explain the many functions of a hereditary molecule: It must be varied, carry a huge amount of information, replicate itself before cell division, and code for traits. To their excitement, as Watson said in his book, *The Double Helix*, DNA's molecular structure turned out to be very "interesting" (see Figure 3.3). Their paper, published in 1953, set off great excitement in the biology community and initiated the explosion of DNA research that continues to this day. Watson, Crick, and Maurice Wilkins (in whose laboratory Franklin had worked) were awarded the Nobel Prize in Medicine in 1962.

Figure 3.2
X–ray diffraction pattern of DNA produced by Rosalind Franklin.

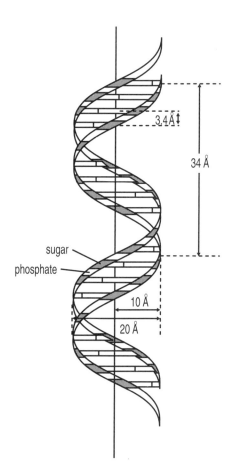

sugar

phosphate

3.4 Å

34 Å

10 Å

20 Å

Figure 3.4
Illustration of a section of the DNA molecule as shown by Watson and Crick in 1953, showing the sugar–phosphate sidepieces and the rungs formed by the four nitrogenous bases.

We now know that DNA is a very long, thin double helix made up of linked nucleotides. Each *nucleotide* consists of one phosphate group, one sugar (deoxyribose) molecule, and one attached nitrogenous base, either adenine, cytosine, guanine, or thymine. The alternating phosphate and sugar molecules provide the sides of the helix much as the sides of a ladder do, and the "rungs" of the helix are made of pairs of nitrogen and carbon-ringed bases (see Figure 3.5). As you determined from Table 3.1, the bases adenine and thymine are present in equal amounts and are joined together; the bases cytosine and guanine are also present in equal amounts and are joined together. Each "rung" is made of a pair of these *complementary* (matching) nitrogenous bases, which are always linked as A with T and C with G, and each of these bases is attached to the sugar molecule in the alternating sugar-phosphate sidepiece or strand. The two bases are held together by hydrogen bonds and, in turn, keep the double strands of the DNA molecule together (see Figures 3.4 and 3.5).

IN THEIR OWN WORDS

The following article is one of the classic papers in scientific literature in the twentieth century. It represents the culmination of forty years of intense scientific research and bridged the gap between earlier discoveries about traits and patterns of inheritance of traits and our modern understanding of how genes code for proteins and how proteins result in traits. In 1953, it was known that DNA was the molecule responsible for

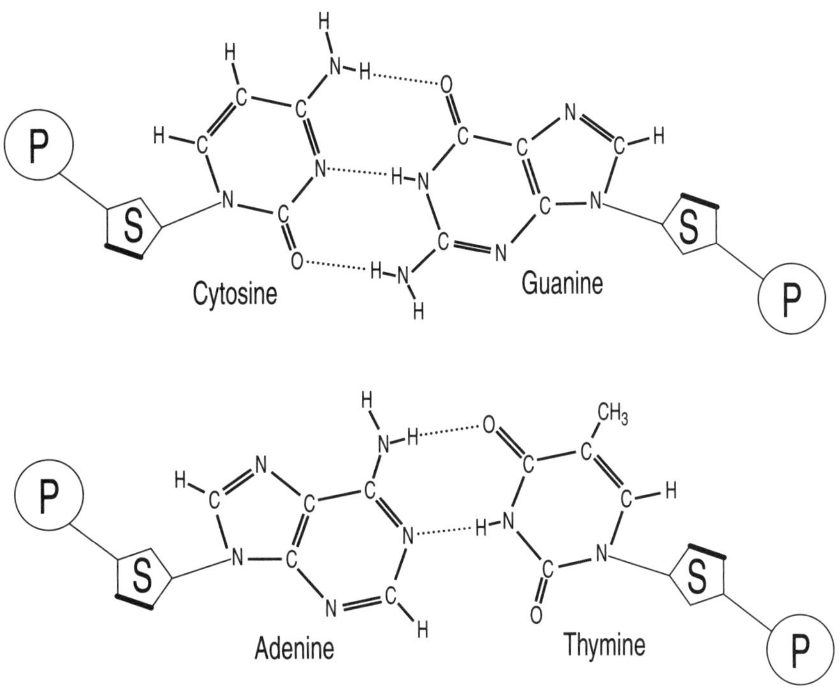

Figure 3.5
Chemical structure and arrangement of the subunits in DNA.

conferring traits within an organism and for the passing on of characteristics from generation to generation. What was not yet known was how this was done. In order to understand how, it was crucial to know the structure of the DNA molecule. In this article, James Watson and Francis Crick proposed a structure for deoxyribonucleic acid, DNA.

As you read, do not be concerned about the complexity of the science or the scientific language. Rather, focus on:

– the description of the structure itself and its composition,

– how the two strands are held together, and

– what Watson and Crick felt was the significance of their model.

Molecular Structure of Nucleic Acids

Reprinted by permission from Nature, volume 171, April 25, 1953, pp. 737-738.
Copyright ©1953 Macmillan Magazines Ltd.

A STRUCTURE FOR DEOXYRIBOSE NUCLEIC ACID

We wish to suggest a structure for the salt of deoxyribose nucleic acid (DNA). This structure has novel features which are of considerable biological interest.

A structure for nucleic acid has already been proposed by Pauling and Corey[1]. They kindly made their manuscript available to us in advance of publication. Their model consists of three intertwined chains, with the phosphates near the fibre axis, and the bases on the outside. In our opinion, this structure is unsatisfactory for two reasons: (1) We believe that the material

Continued on next page

which gives the X-ray diagrams is the salt, not the free acid. Without the acidic hydrogen atoms it is not clear what forces would hold the structure together, especially as the negatively charged phosphates near the axis will repel each other. (2) Some of the van der Waals distances appear to be too small.

Another three-chain structure has also been suggested by Fraser (in the press). In his model the phosphates are on the outside and the bases on the inside, linked together by hydrogen bonds. This structure as described is rather ill-defined, and for this reason we shall not comment on it.

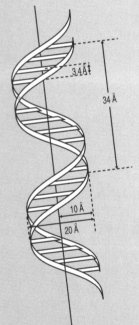

This figure is purely diagrammatic. The two ribbons symbolize the two phosphate-sugar chains, and the horizontal rods [symbolize] the pairs of bases holding the chains together. The vertical line marks the axis.

We wish to put forward a radically different structure for the salt of deoxyribose nucleic acid. This structure has two helical chains each coiled round the same axis (see diagram).

We have made the usual chemical assumptions, namely, that each chain consists of phosphate diester groups joining ß-d-deoxyribofuranose residues with 3′,5′ linkages. The two chains (but not their bases) are related by a dyad perpendicular to the fibre axis. Both chains follow right-handed helices, but owing to the dyad the sequences of the atoms in the two chains run in opposite directions. Each chain loosely resembles Furberg's[2] model No. 1; that is, the bases are on the inside of the helix and the phosphates on the outside. The configuration of the sugar and the atoms near it is close to Furberg's 'standard configuration', the sugar being roughly perpendicular to the attached base. There is a residue on each chain every 3.4 Å [angstroms] in the z-direction. We have assumed an angle of 36° between adjacent residues in the same chain, so that the structure repeats after 10 residues on each chain, that is, after 34 Å. The distance of a phosphorus atom from the fibre axis is 10 Å. As the phosphates are on the outside, cations have easy access to them.

The structure is an open one, and its water content is rather high. At lower water contents we would expect the bases

to tilt so that the structure could become more compact. The novel feature of the structure is the manner in which the two chains are held together by the purine and pyrimidine bases. The planes of the bases are perpendicular to the fibre axis. They are joined together in pairs, a single base from one chain being hydrogen-bonded to a single base from the other chain, so that the two lie side by side with identical z-coordinates. One of the pair must be a purine and the other a pyrimidine for bonding to occur. The hydrogen bonds are made as follows: purine position 1 to pyrimidine position 1; purine position 6 to pyrimidine position 6.

If it is assumed that the bases only occur in the structure in the most plausible tautomeric forms (that is, with the keto rather than the enol configurations) it is found that only specific pairs of bases can bond together. These pairs are: adenine (purine) with thymine (pyrimidine), and guanine (purine) with cytosine (pyrimidine).

In other words, if an adenine forms one member of a pair, on either chain, then on these assumptions the other member must be thymine; similarly for guanine and cytosine. The sequence of bases on a single chain does not appear to be restricted in any way. However, if only specific pairs of bases can be formed, it follows that if the sequence of bases on one chain is given, then the sequence on the other chain is automatically determined.

It has been found experimentally[3,4] that the ratio of the amounts of adenine to thymine, and the ratio of guanine to cytosine, are always very close to unity for deoxyribose nucleic acid.

It is probably impossible to build this structure with a ribose sugar in place of the deoxyribose, as the extra oxygen atom would make too close a van der Waals contact.

The previously published X-ray data[5,6] on deoxyribose nucleic acid are insufficient for a rigorous test of our structure. So far as we can tell, it is roughly compatible with the experimental data, but it must be regarded as unproved until it has been checked against more exact results. Some of these are given in the following communications. We were not aware of the details of the results presented there when we devised our structure, which rests mainly though not entirely on published experimental data and stereo-chemical arguments.

It has not escaped our notice that the specific pairing we have postulated immediately suggests a possible copying mechanism for the genetic material.

Full details of the structure, including the conditions assumed in building it, together with a set of coordinates for the atoms, will be published elsewhere.

We are much indebted to Dr. Jerry Donohue for constant advice and criticism, especially on interatomic distances. We have also been stimulated by a knowledge of the general nature of the unpublished experimental results and ideas of Dr. M. H. F. Wilkins, Dr. R. E. Franklin and their co-workers at King's College, London. One of us

(J. D. W.) has been aided by a fellowship from the National Foundation for Infantile Paralysis.
J. D. Watson
F. H. C. Crick

Medical Research Council Unit for the Study of the Molecular Structure of Biological Systems
Cavendish Laboratory,
Cambridge.
April 2.

NOTES
1. Pauling, L., and Corey, R. B., Nature, 171, 346 (1953); Proc. U.S. Nat. Acad. Sci., 39, 84 (1953).
2. Furberg, S., Acta Chem. Scand., 6, 634 (1952).
3. Chargaff, E., for references see Zamenhof, S., Brawerman, G., and Chargaff, E., Biochim. et Biophys. Acta, 9, 402 (1952).
4. Wyatt, G. R., J. Gen. Physiol., 36, 201 (1952).
5. Astbury, W. T., Symp. Soc. Exp. Biol. 1, Nucleic Acid, 66 (Camb. Univ. Press, 1947).
6. Wilkins, M. H. F., and Randall, J. T., Biochim. et Biophys. Acta, 10, 192 (1953).

DECIPHERING THE CODE

The intriguing questions now became: How can such a simple molecule, made of only four nitrogenous bases or "letters," be the basis of all the traits we see in living things? What is the code these "letters" make? How can this code translate into heritable traits? Since round peas are the result of the action of an enzyme (protein), there must be a relationship between the sequence of the nucleotides (the code) and the production or synthesis of protein.

It had been hypothesized that *messenger ribonucleic acid (mRNA)*, which is found both in the nucleus and throughout the cytoplasm, might be involved in making protein. *Ribonucleic acid (RNA)* is a single-strand molecule made of one sidepiece of alternating sugar *(ribose)* and phosphate groups. Its unpaired nitrogenous bases, which are attached to the sugar, are: adenine (A), cytosine (C), and guanine (G); *uracil* (U), a pyrimidine similar to thymine, pairs with adenine. In 1961, Marshall

Nirenberg tested the hypothesis that RNA was involved in making protein. Using a mixture made from cells (cell extract) that was capable of making protein, he demonstrated that in the absence of RNA, no proteins were made; but when he added RNA, proteins were made. He then made an RNA molecule composed entirely of one kind of nitrogenous base, uracil. When he added this RNA to a cellular extract, he found it made a protein containing only one kind of amino acid, phenylalanine.

Nirenberg concluded that U-U-U coded for phenylalanine. Using this method, the codes for each of the amino acids were found. Further experiments showed that the code in DNA is a triplet, that is, it requires three nucleotides (called a *codon*) to transcribe the DNA code to messenger RNA (mRNA) which then translates the code into an amino acid. Soon, all the codes for the 20 amino acids were deciphered (see Figure 3.6).

The connections among DNA, genes, and proteins had begun in 1941 with a classic experiment by George Beadle and Edward Tatum in which they determined that a gene in an organism contains the information to synthesize a protein. Further research on the mechanics of how the DNA blueprint is transcribed into the mRNA code and how the mRNA code is translated into a protein sharpened the definition of a gene to mean a segment of the DNA molecule that codes for an enzyme or protein.

Figure 3.6

Three nitrogenous bases (a codon) code for a specific amino acid.

codons	GCA GCC GCG GCU	AGA AGG CGA CGC CGG CGU	AAC AAU	GAC GAU	UGC UGU	CAA CAG	GAA GAG
abbreviations for amino acids	ala	arg	asn	asp	cys	gln	glu

codons	GGA GGC GGG GGU	CAC CAU	AUA AUC AUU	UUA UUG CUA CUC CUG CUU	AAA AAG	AUG	UUC UUU
abbreviations for amino acids	gly	his	ile	leu	lys	met	phe

codons	CCA CCC CCG CCU	AGC AGU UCA UCC UCG UCU	ACA ACC ACG ACU	UGG	UAC UAU	GUA GUC GUG GUU	UAA UAG UGA
abbreviations for amino acids	pro	ser	thr	trp	tyr	val	stop

For ease of writing, the amino acids are abbreviated as follows:

ala	alanine	leu	leucine
arg	arginine	lys	lysine
asn	asparagine	met	methionine
asp	aspartic acid	phe	phenylalanine
cys	cysteine	pro	proline
gln	glutamine	ser	serine
glu	glutamic acid	thr	threonine
gly	glycine	trp	tryptophan
his	histidine	tyr	tyrosine
ile	isoleucine	val	valine

From DNA to RNA to Protein

INTRODUCTION In Learning Experience 2—No Matter What Your Shape, you found that the shape of peas, round or wrinkled, is determined by whether a single enzyme, starch-branching enzyme (SBEI), is functioning correctly. Similarly, whether an individual suffers from sickle cell disorder depends on whether he or she has normal or sickling hemoglobin protein. Where does a pea or a human get the information for making proteins? In the following activities, you will investigate DNA as the molecule which contains all the information for the traits of organisms and determine how this information is transferred into protein.

After transcription, the mRNA leaves the nucleus, moves into the cytoplasm, and binds to a structure called a *ribosome,* which is made up of protein and is a third kind of ribonucleic acid (rRNA). That is where the actual translation to protein takes place. The mRNA moves along the ribosome, each triplet codon is shown or "exposed," and yet another type of RNA molecule called a *transfer RNA (tRNA)* interacts with the codon. At one end of this tRNA are three exposed nitrogenous bases (the *anticodon*) that are complementary to the mRNA codon exposed. At the other end of the tRNA is the specific amino acid that is encoded by the codon. The tRNA binds to the mRNA by matching nucleotides (C–G, A–U, etc.) and the amino acid attaches to the preceding amino acid, forming a growing chain of the *polypeptide* (see Figure 3.7).

In this activity, you will use pushpins to represent the 18 nitrogenous bases of a DNA sequence. You will be modeling the processes of *transcription* (DNA—> mRNA) and *translation* (mRNA —> amino acid chain or protein).

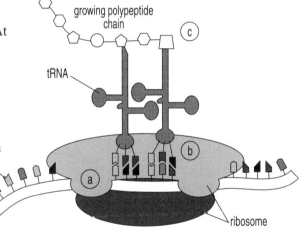

Figure 3.7
Translation process. The mRNA binds to the ribosome and exposes a codon (a). A tRNA with a complementary sequence binds to the mRNA codon (b) and adds the amino acid it is carrying to the growing polypeptide chain (c).

▶ MATERIALS NEEDED

For each group of four students:
- 54 pushpins in assorted colors
- 4 strips of corrugated cardboard (24 cm x 4 cm)
- 1 metric ruler
- 1 felt-tip marker
- 1 scissors or scalpel
- masking tape

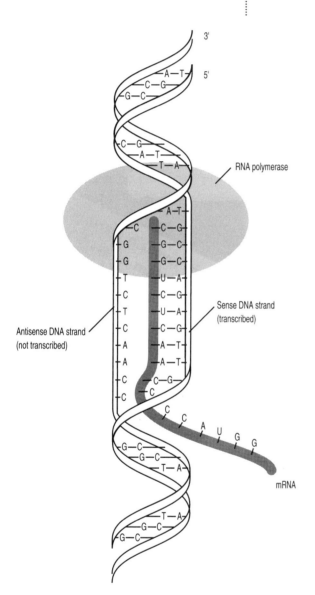

Figure 3.8
Representation of transcription showing DNA "unzipping." RNA nucleotides form a new strand of mRNA with the aid of RNA polymerase, an enzyme.

▶ PROCEDURE

1. Select 18 pushpins to represent the nitrogenous bases of the DNA sequence that follows in colors that correspond to the color key shown. For example, for this sequence you will need five red pushpins to represent the five adenines.

    Use the following color key to represent the bases:

red pushpin	=	adenine or A
blue pushpin	=	guanine or G
yellow pushpin	=	cytosine or C
green pushpin	=	thymine or T
clear pushpin	=	uracil or U

    The following is a sequence of a portion of a DNA molecule:

    TACCACGTGGACTGAGGA

2. Arrange these pushpins vertically according to the above DNA sequence along the center of one of the long cardboard strips. This is your "sense" strand. On the left of each pushpin place the letter S to represent the sugar that is attached to each nitrogenous base. Place a letter P between each "S." The arrangement of the "P" and "S" illustrate the alternating sugar-phosphate sidepiece.

3. To the right of each pushpin, write on the cardboard the letter of the complementary base in the second strand of this DNA segment. Refer to Figure 3.5 for how nitrogenous bases pair. (This written strand is the "antisense" DNA strand and will not be used in this simulation.) Label this cardboard strip "DNA."

4. Place a second cardboard strip next to your "sense" strand of DNA pushpins. Use this strand as the code to make a complementary mRNA strand. You may first wish to write out the mRNA sequence strand on paper. This process, called transcription, occurs in the nucleus of the cell (see Figure 3.8). Obtain 18 more pushpins in the appropriate key colors and place them in order along the strip.

5. On the cardboard strip of your mRNA sequence, place the letter S to the left of each pushpin and the letter P between each "S," as you did in step 2. Label this cardboard strip "mRNA." Use a marker and draw a line

under every three nucleotides (base, sugar, and phosphate). For example, your first underline would be the codon AUG.

6. Model what happens after transcription by moving the mRNA cardboard strip away from the DNA strip (to simulate that the mRNA is moving from the nucleus to the cytoplasm).

7. Measure and cut the third cardboard strip into six equal sections (each 4 cm x 4 cm). Place the sections along the mRNA strand. On each of these sections place pushpins representing the complementary tRNA anticodon. For example, the first anticodon would be UAC, attached to the AUG codon.

8. Measure and cut the fourth cardboard strip into six equal sections (each now 4 cm x 4 cm). Use the genetic code table in Figure 3.6 to determine the amino acid for each of the six triplet mRNA codons. With your marker write the name of each amino acid on a separate cardboard section.

9. Line the amino acids next to the appropriate tRNA anticodon. Join the six amino acids together with tape. This series of amino acids is a polypeptide chain or protein.

NOTE: Remember from your reading "Deciphering the Code" in "The Molecule of Life" that every triplet nucleotide is a codon. This codon determines which tRNA will bind to the codon and, therefore, which amino acid will be added.

NOTE: It is the triplet codon AUG that codes for the amino acid methionine. Methionine is the first amino acid of a protein chain and is considered the "start" codon. tRNA is the carrier of the amino acid to the mRNA strand.

▶ ANALYSIS

Consult with your group and write responses to the following in your notebook.

1. Explain in your own words how information in DNA codes for a protein.

2. How might sickling hemoglobin occur in an individual?

3. Why do you think DNA does not code directly but uses mRNA?

4. The DNA code you used is the first 18 nucleotides for ß-hemoglobin. With this information and the knowledge you have gained from the readings and the activity you have just completed, define a "gene."

THE DNA SHUFFLE

INTRODUCTION   In this activity you will role-play the complete process of protein synthesis. You and your classmates will take on three roles: codons (mRNA triplet bases), anticodons (tRNA triplet bases), and amino acids. The Procedure describes the process in detail. It is important to understand each step, not just your role, because these processes are essential to all living things. The information stored in DNA is transcribed to mRNA, which codes for the assembling of a polypeptide chain; and the actions of proteins determine the traits or characteristics of all organisms.

ACTIVITY

▶ MATERIALS NEEDED

For the class:
- DNA sequence paper model
- index cards representing codons, anticodons, and amino acids
- tape

▶ PROCEDURE

1. Read through the entire Procedure before beginning the role-play.

2. Take one index card from the stack your teacher has prepared. Read what is on the card and identify to which of the above groups it belongs. When everyone has a card, form three groups of students (codons, anticodons, and amino acids).

3. Look at the gene (DNA sequence model) taped to the board. Determine who has the mRNA card that matches the first triplet on the gene. That person should line up by the board in front of that triplet sequence.

4. Repeat step 3 for each codon. When all codons are in order, the process of transcription is complete.

5. Students representing the mRNA codon strand move away from the DNA to the center of the room (or other designated open area).

6. Students in the remaining two groups now pair up; that is, each person with an anticodon card, CAC for example, pairs up with the matching person from the amino acid group (valine).

7. These pairs move to the mRNA strand (of students). The anticodon attaches to the appropriate codon.

8. One by one, the anticodons move away from the mRNA strand and go to one side of the room. As they do, each amino acid joins hands with the preceding amino acid.

9. This amino acid chain is called a polypeptide chain or a protein. As a group, this protein should move away from the mRNA strand to another part of the room. This completes the process of translation.

The Wonderful Mistake

from "The Wonderful Mistake" copyright ©1979 by Lewis Thomas from *The Medusa and the Snail,* by Lewis Thomas pp. 22–24. Used by permission of Viking Penguin, a division of Penguin Books USA Inc.

The greatest single achievement of nature to date was surely the invention of the molecule of DNA. We have had it from the very beginning, built into the first cell to emerge, membranes and all, somewhere in the soupy water of the cooling planet three thousand million years or so ago. All of today's DNA, strung through all the cells of the earth, is simply an extension and elaboration of that first molecule. In a fundamental sense we cannot claim to have made progress, since the method used for growth and replication is essentially unchanged.

But we have made progress in all kinds of other ways. Although it is out of fashion today to talk of progress in evolution if you use that word to mean anything like improvement, implying some sort of value judgment beyond the reach of science, I cannot think of a better term to describe what has happened. After all, to have come all the way from a system of life possessing only one kind of primitive microbial cell, living out colorless lives in hummocks of algal mats, to what we see around us today—the City of Paris, the State of Iowa, Cambridge University, Woods Hole, the succession of travertine-lined waterfalls and lakes like flights of great stairs in Yugoslavia's Plitvice, the horse-chestnut tree in my backyard, and the columns of neurones arranged in modules in the cerebral cortex of vertebrates—*has* to represent improvement. We have come a long way on that old molecule.

We could never have done it with human intelligence, even if molecular biologists had been flown in by satellite at the beginning, laboratories and all, from some other solar system. We have evolved scientists, to be sure, and so we know a lot about DNA, but if our kind of mind had been confronted with the problem of designing a similar replicating molecule, starting from scratch, we'd never have succeeded. We would have made one fatal mistake: our molecule would have been perfect. Given enough time, we would have figured out how to do this, nucleotides, enzymes, and all, to make flawless, exact copies, but it would never have occurred to us, thinking as we do, that the thing had to be able to make errors.

The capacity to blunder slightly is the real marvel of DNA. Without this special attribute, we would still be anaerobic bacteria and there would be no music. Viewed individually, one by one, each of the mutations that have brought us along represents a random, totally spontaneous accident, but it is no accident at all that mutations occur; the molecule of DNA was ordained from the beginning to make small mistakes.

If we had been doing it, we would have found some way to correct this, and evolution would have been stopped in its tracks. Imagine the consternation of human scientists, successfully engaged in the letter-perfect replication of prokaryotes, nonnucleated cells like bacteria, when nucleated cells suddenly turned up. Think of the agitated commissions assembled to explain the scandalous proliferations of trilobites all over the place, the mass firings, the withdrawal of tenure.

To err is human, we say, but we don't like the idea much, and it is harder still to accept the fact that erring is biological as well. We prefer sticking to the point, and insuring ourselves against change. But there it is: we are here by the purest chance, and by mistake at that. Somewhere along the line, nucleotides were edged apart to let new ones in; maybe viruses moved in, carrying along bits of other, foreign genomes; radiation from the sun or from outer-

Continued on next page

space caused tiny cracks in the molecule, and humanity was conceived.

And maybe, given the fundamental instability of the molecule, it had to turn out this way. After all, if you have a mechanism designed to keep changing the ways of living, and if all the new forms have to fit together as they plainly do, and if every improvised new gene representing an embellishment in an individual is likely to be selected for the species, and if you have enough time, maybe the system is simply bound to develop brains sooner or later, and awareness.

Biology needs a better word than "error" for the driving force in evolution. Or maybe "error" will do after all, when you remember that it came from an old root meaning to wander about, looking for something.

EXTENDING IDEAS

▶ Read *The Double Helix* by James Watson or *Rosalind Franklin* by Anne Sayre and write an essay describing your view of how a major discovery was made. Include the nature of the approach, the importance of collaborative work, and the roles that luck or imagination might have played.

ON THE JOB

CRIMINALIST   The courtroom was completely silent. The most important piece of evidence linking Jason Benton to the attempted murder of a state university student had been placed before the jury. That evidence came from forensic science, the practical application of science and medicine to law.

On a warm day in autumn, Missy, a freshman in college, was at a campus party with her roommate and decided to leave early and walk back to the dormitory alone. When her roommate returned, it was obvious that Missy had not made it back. After retracing the path that Missy was assumed to have taken home, her roommate and other friends found her lying just off the road, obviously terrified and suffering from severe trauma. She was coherent and said that she had been mugged and beaten on her way home. When police arrived, she described that her attacker had grabbed her from behind. She never clearly saw his face but she recalled that he had spoken with an angry stutter and a foreign-sounding accent. Because of the stutter and accent, the police suspected a well-known, recently paroled criminal, Jason Benton, and they picked him up for questioning.

A forensics team that included the state's senior criminalist immediately studied the area for evidence, finding little. But, Missy had scratched her attacker and got some of his skin beneath her fingernails. It was just the break the forensics team needed. The skin scrapings were carefully extracted and sent to a nearby forensics lab, along with all of the other physical evidence. There, criminalists began DNA testing on the tissue sample.

The scientific principle behind the DNA test is that each person (except for identical twins) has his or her own unique DNA "fingerprint." If the DNA in the tissue Missy scratched from her attacker matched that of Jason Benton's blood, he would more than likely be found guilty. With a suspect's life at stake, it is vital to have accurate tests, and to avoid any contamination of the samples by other DNA. When the case went to trial, the jury would want to be positive that the DNA test was reliable.

Since the early 1980s, DNA analysis has been the most reliable tool for identifying criminals of violent crimes, particularly sexual assaults. DNA can also be used to identify serial criminals, prove paternity, and identify human remains.

DNA is just one of the many substances criminalists work with. They must have a knowledge of crime scene investigations, trace evidence examinations, physiological fluid analysis (including blood, urine, semen, and saliva), controlled substance analysis, and blood alcohol analysis. When evidence from a crime comes to the laboratory, criminalists must be able to decide what tests should be done. They must be able to conduct analyses or examinations on hair, fiber, soil, paint, glass, building materials, projectiles, and footwear. Once all the evidence has been tested and examined, it must be presented to local law enforcement personnel for possible use in a trial.

There are many different routes to becoming a criminalist. Colleges offer majors in biology and chemistry, and some universities have more specialized majors in forensic science. Once a bachelor's degree is acquired, a student can enter the field or continue his/her education through the doctoral level.

Back in the courtroom, the DNA evidence was presented to the jury. The DNA "fingerprints" from the tissue and from Jason Benton's blood matched perfectly. The defense then tried to show that the criminalists had contaminated the samples by careless handling, but the work had been done with the utmost care. After 10 hours of deliberating, passing around photographs of the crime scene, dissecting transcripts, and examining the DNA evidence, the jury delivered their verdict: "We, the jury, hereby find Jason Benton guilty of assault and attempted murder."

CHANGE IN SEQUENCE, CHANGE IN TRAIT

PROLOGUE

Excerpted from "Why So Many Errors in Our DNA?" Blazing a Genetic Trail, *by Maya Pines, Howard Hughes Medical Institute, 1991, p. 17.*

*A*s scientists learn to read the instructions in our genes, they are discovering that much of our DNA is riddled with errors.

Fortunately, most of these errors are harmless. Considering the difficulties involved—the 6 feet of DNA in a human cell consists of 6 billion subunits, or base pairs, coiled and tightly packed into 46 chromosomes, all of which must be duplicated every time a cell divides—our general state of health is something of a miracle.

We each inherit hundreds of genetic mutations from our parents, as they did from their forebears. In addition, the DNA in our own cells undergoes an estimated 30 new mutations during our lifetime, either through mistakes during DNA copying or cell division or, more often, because of damage from the environment. Bits of our DNA may be deleted, inserted, broken, or substituted. But most of these changes affect only the parts of DNA that do not contain a gene's instructions, so we need not worry about them.

Mutations are changes which occur in the sequence of DNA of an organism. In this learning experience, you will explore what happens when different kinds of mutations occur in the sequence of a gene and how these changes are reflected in the traits determined by that gene.

SIMPLE MISTAKE, SERIOUS CONSEQUENCES

INTRODUCTION   In Learning Experience 2—No Matter What Your Shape, Denzel learned that a single change in an amino acid in the hemoglobin protein can mean the difference between being healthy and having a life-threatening disease. He knows that proteins were made up of amino acid subunits and that the kind and arrangement of amino acids in a protein determined its function. But he did not know how changes in a protein could occur.

In the following activity, you will determine how a change or mutation in the DNA sequence that codes for hemoglobin can cause a change in the protein, resulting in sickling hemoglobin.

▶ TASK

1. Obtain a copy from your teacher of the mRNA sequences of:
    – normal hemoglobin (hemoglobin A)
    – sickling hemoglobin (hemoglobin S)

2. Draw a line under every three nucleotides. For example, your first underline would be AUG. This is the first codon. Complete for the sequence of both normal and sickling hemoglobin.

3. Using the codon table in Figure 3.6, determine the amino acid sequence of the protein encoded in each RNA triplet codon. List the amino acids in each sequence in your notebook.

▶ ANALYSIS

Write responses to the following in your notebook.

1. What is the difference between the mRNA sequences for normal and sickling hemoglobin? How does this affect the proteins?

2. What effect does the change have on the function of the proteins? (Use information from Learning Experience 2, if necessary.)

3. How does this change affect the individual who carries the DNA sequence for hemoglobin S (sickling)?

4. Using the two mRNA sequences as guides, write the sequences for each of the strands of normal and sickling DNA. What difference do you note?

5. Explain in a short paragraph the relationships among the following: DNA, protein, and a trait.

THE ERROR OF OUR DNA

INTRODUCTION   As you read in the Lewis Thomas article, many errors in DNA sequence occur during the process of DNA replication. Some mutations occur because of environmental influences. (Radiation and certain chemicals can cause changes in the bases in DNA.) Most mutations occur at random or by chance and in regions which do not contain the instructions for encoding a protein. Problems only arise when an error in DNA alters the information needed to synthesize a functional protein. A mutation may involve the deletion, insertion, or duplication of a portion of a DNA molecule, or the substitution of one or more nucleotides in the molecule.

In this activity you will alter a DNA sequence and describe how that error may affect the protein the DNA is coding for.

▶ TASK

1. Write the following DNA sequence in your notebook.

    TACCGTCTGAAAGGT

2. Transcribe the DNA sequence into RNA and then, using Figure 3.6, translate the sequence into protein.

3. Depending on your group number, alter the DNA sequence from Task step 1 as follows:
    –   If your group number is 1, substitute one base for another.
    –   If your group number is 2, insert one new base.
    –   If your group number is 3, insert two new bases.
    –   If your group number is 4, insert three new bases.
    –   If your group number is 5, delete one base.
    –   If your group number is 6, delete two bases.
    –   If your group number is 7, delete three bases.

4. Transcribe the new sequence into mRNA, then translate the resulting sequence into protein. Be prepared to tell the class the results of your particular mutation: Describe what your group did, how that altered the amino acid sequence, and what implications this may have.

▶ ANALYSIS

Discuss responses to the following with your group.

1. When might a change in the DNA sequence have no effect on the traits of an organism?

2. Where would a change in the DNA sequence make the greatest difference in the protein sequence?

3. The Prologue quotation suggests that the DNA in our own cells undergoes an estimated 30 mutations during our lifetime. What might be the results of these mutations?

4. How do you think these changes might affect your life? Will these changes be passed on to your children? Explain your response.

LIVING WITH GAUCHER'S DISEASE

Gaucher's disease (pronounced go-shayz) is caused by mutations in the gene that codes for the enzyme glucocerebrosidase (GC), a protein that normally breaks down a fatty substance called glucosylceramide. In the disease, the mutated GC gene produces a nonfunctional enzyme. As a result, the fatty substance accumulates in the liver, spleen, and bone marrow, and on rare occasions, the brain.

Typical symptoms of Gaucher's disease include an enlarged spleen and/or liver, bone deterioration, and loss of bone density with multiple fractures and "bone crisis" (bone pain). In some patients, there is also progressive nervous system degeneration. Other symptoms include general fatigue, decreased ability to provide oxygen to the blood, disruption of kidney functions, and increased bleeding. Most people do not develop all of the possible symptoms, and severity of the disease varies enormously.

Fewer than one in 40,000 people in the general population have Gaucher's disease. The incidence is significantly higher among Jews of Eastern European descent. The higher frequency of this disease among this population has led to the mistaken notion that this disease is a "Jewish disease." In fact, individuals of any ethnic or racial background may be affected. The following case study describes one person's experience with the disease.

Sandy's Story: My Spleen Was Now Enormous

Source: Gauchers News, September 1995 © Gauchers Association, 25 West Cottages, London, NW6 1RJ, UK.

I was diagnosed at the age of four at St. Thomas's Hospital in London after numerous blood tests and a bone marrow sample. My parents were told that it was a very rare disease and there was no cure or treatment available apart from removing my spleen if it became too enlarged. I come from a non-Jewish family of five girls—the disease was found in three of us.

Apart from nosebleeds, tiredness causing me to fall asleep in school and aching legs, it didn't present much of a problem for me until the age of 13 when I suffered what is now known as a bone crisis, but at that stage it was a mystery as nothing showed up on the X-rays. I was given 3 weeks' traction in an orthopaedic hospital. This occurred again at the age of 18 but this time 2 weeks' bed rest was the advice and it worked.

BONE MARROW TEST

I was again reviewed by the local haematologist who repeated the bone marrow test in my chest. This is not a pleasant experience and it would now appear that it is not even necessary. I was told although my spleen was enlarged, I should keep it as long as possible as evidence showed that once the spleen was gone, my bones would degenerate more rapidly.

I gave birth to my son at the age of 23 without too many problems. My spleen had enlarged quite a lot during the pregnancy but I was told that there was no reason why I should not have more children. Six weeks before the birth of my daughter, I was taken into the hospital for total bed rest as again it seemed that the pregnancy had accelerated the growth of my spleen and it was possible that it may rupture. I was now told no more children as it would be far too dangerous.

From this time on I continuously looked pregnant. I never had a waistline in the past but it was even more pronounced by my very large spleen. . .

TIME TO FIND OUT

In 1991 I decided that it was about time I found out about Gauchers disease. A friend saw an article in a national Sunday paper about a young boy who had this rare disease and was flying backwards and forwards to the USA to receive a life saving drug. She recognised the name of the disease and gave me the article.

After having the disease for all these years and meeting many different doctors who have never heard of it, apart from in their reference books, here it was in the papers, another person with Gauchers. This was my opportunity to follow it up and enquire about the treatment and see if I could be of help towards any research into it. I contacted the newspaper editor who . . . gave me the address of the National Gaucher Foundation in America who I wrote to immediately. They contacted me and asked if I would be interested in taking part in trials [that may lead to the possible cure of the disease.] . . .

Until recently, patient care and therapy for Gaucher's disease was directed at managing (that is, relieving) the symptoms. Depending on symptoms, therapy includes the following measures, either alone or in combination: bed rest, anti-inflammatory medicines for acute pain, biofeedback techniques for pain management, hyperbaric oxygen therapy for the treatment of bone crisis, splenectomy (removal of the spleen), and oxygen therapy. None of these approaches is totally satisfactory. Spleen removal increases the susceptibility to bacterial disease, and may lead to increased liver and bone symptoms.

Recently researchers have made progress in the development of treatments that go beyond dealing with symptoms. Modified or variant

GC (glucocerebrosidase) enzyme has been evaluated in clinical trials which showed that repeated infusions of the enzyme reduced the signs and symptoms of the disease, and reversed the disease progression. The modified enzyme called Ceredase is believed to be the first true therapeutic breakthrough. The administration of Ceredase will be required at regular intervals (usually several times a week as an infusion) throughout an individual's lifetime. Ceredase costs $150,000 a year for each patient and would be an effective therapy for those who can get it, but not a cure for the underlying disease. (The cost is usually covered by health insurance.)

Researchers are also pursuing avenues of genetic investigations that may point to a possible cure for the disease. Efforts are underway to develop ways to introduce normal genes for glucocerebrosidase into cells of the affected person. These cells would then produce sufficient normal amounts of active GC. This approach is highly experimental at present, and technical hurdles have to be overcome to demonstrate that it is safe.

▶ ANALYSIS

Write responses to the following in your notebook.

1. Should Sandy go to America for treatment? List three reasons why she should go and three reasons why she should not go.

2. Describe the consequences of each of the reasons you listed above.

3. If Sandy were to be part of this test, what should the doctors tell her about the treatment?

4. Should Sandy sign a permission form that says she will not sue the doctors if the treatment harms her in some way? Why or why not?

5. In a short paragraph, explain what choice you would make if you were Sandy. Include in your paragraph who should pay for the treatments you chose to get.

EXTENDING IDEAS

Great concern is being expressed in the media about the rise in "environmentally induced" cancers and the role of mutagens, such as chemicals and radiation, in the environment in causing cancer. Certain of these chemicals are known to be mutagens, that is, chemicals known to cause mutations.

Identify three mutagens that are known to be carcinogens (agents known to cause cancer), and research the relationship between carcinogens and mutagens.

ON THE JOB

LOBBYIST   Jay is preparing for his afternoon appointment with one of his state's representatives in Congress. He is reading literature and background data about the latest legislation regarding health, and about life insurance companies' access to results of testing for genetic disorders. Jay was hired by a well-known insurance provider; the company hopes Jay will be able to convince members of the U.S. Congress that companies dealing in life and health insurance must have access to these records in order to protect themselves against high-risk consumers. This extra protection will allow them to keep rates affordable for their lower-risk customers, thus allowing more people to acquire better coverage. Jay will try to meet with numerous members of Congress, pass out literature to people having political influence, and keep one step ahead of the latest information regarding genetics testing. He needs to come across as believable and knowledgeable in the field to the politicians he will try to sway. This may involve hours of research, not only on genetics testing but on the insurance industry, on general medicine, and on the running of government.

But, Jay's job includes more than just being concerned with genetics testing. He has to be aware of all issues that might somehow affect the company for which he works. One such issue is socialized medicine as practiced in areas such as Great Britain and Scandinavia, where the governments control health insurance as well as health services.

At the same time, Christine is making similar appointments and reading similar reports. She is a lobbyist for a small special-interest group that strongly opposes genetics-testing results being made available to insurance companies. Many in this group feel that some people will be unfairly penalized with higher premiums because of hereditary diseases that may never appear, or even be turned down completely for insurance. This group is using the right to privacy as their main argument. Christine has collected data from a couple who had a genetics test for Down Syndrome on their unborn child. Even though that child was born with only a very slight case of the syndrome, the parents have had to pay a much higher premium than is usual, a premium far above what they can afford.

Jay and Christine have similar jobs, and interests in the same issues, but with hopes for opposite outcomes. Jay first learned about lobbying in his political science class in high school. He majored in political science in college, and got most of his education in genetics after being hired by the insurance company.

Christine, on the other hand, has a good background in genetics from her college classes. After taking a trip to Washington, D.C., during her senior year, she became interested in government and lobbying, and researched companies and nonprofit organizations that might be able to use her services. She also spent time researching the political process, finding out who influences legislation and how different members of Congress have voted in the past on medical issues.

Jay and Christine pass each other often in the halls of Congress, as lobbyists play a large role in the decisions made by our government. They are on opposite sides of most issues. By informing congresspersons about issues from different perspectives, they help busy legislators make informed choices among complex alternatives.

A Gene for Everything and Every Gene in Its Place

PROLOGUE In 1820, long before DNA became a household acronym and scientists understood the biochemistry of genes, an Augustinian monk, Gregor Mendel, theorized that traits in peas were the result of distinct, heritable entities he called "factors." For many years, Mendel's theory of factors was disregarded and ignored. However, in the early 1900s many researchers, such as Hugo de Vries of Holland, Carl Correns of Germany and Erich von Tschermak of Austria, realized that Mendel's factors could be used to explain many of their own observations about the ways in which traits were inherited in other plants.

While the concept of a gene as a unit of inheritance was acknowledged at this time, little was known about its physical nature. In 1902, Walter Sutton, a scientist at Columbia University in New York City, observed structures arranged in patterns in the nuclei of grasshopper sperm cells. He felt that these patterns could explain much of the data on the inheritance of characteristics. His discovery of these structures, chromosomes, laid the foundation for fundamental research into the molecular basis of genetics. In the following two decades, the structure of chromosomes was determined.

Between 1915 and 1926, Thomas Hunt Morgan, an American zoologist, published two books, *The Mechanism of Mendelian Heredity* and *The Theory of the Gene*, in which he supported Mendel's idea of a gene as a unit responsible for traits which could be inherited. He further extended the concept by concluding that all the genes of an organism existed on chromosomes and that the patterns in which traits were inherited could be explained by how genes were connected to one another on these chromosomes.

All these scientists conducted their research prior to Watson and Crick's discovery of DNA. In this learning experience, you will be exploring the structure of the chromosome, using your knowledge about DNA and genes, in order to determine the mechanisms by which traits are inherited.

DNA, DNA, WHEREFORE ART THOU, DNA?

INTRODUCTION   What is a chromosome, what is it made of, and what is the relationship of a chromosome to DNA and genes? What was Sutton seeing when he first made his critical observation? Looking at a chromosome under a microscope is a little like looking at a building from the street. When you view a building from the outside you see only the external structure, the shape of the building, and any parts which make up the outside components: windows, doors, brickwork, and decorations. A cross section of that same building would give you a very different perspective. It would reveal to you what the interior of the building was made of, how it was put together, and what kind of activities were going on inside.

If you were to use a microscope to look at a chromosome that had been stained with a dye, you would see a striped rod-shaped structure (see Figure 5.1). This first view might not tell you much about a chromosome and, as with a building, you would have to look deeper to determine exactly what it was made of, how it was put together, and what kind of activities were going on.

Chromosomes, found in the nucleus of cells, are composed of long strands of DNA tightly packed with proteins. The DNA within the chromosome is highly coiled and entwined with protein structures called *histones.* The chromosome also contains other proteins which are involved in transcribing and regulating the expression of the information in the DNA. The stripes or banding patterns that are observed when a chromosome is stained are the result of the dyes binding to specific types of proteins. Each chromosome pair in an organism has a unique banding pattern.

Emerging from DNA within a chromosome is the RNA that is being transcribed from genes on the chromosome (see Figure 5.2).

In the following activity, you will construct a chromosome and examine the relationship between genes and chromosomes.

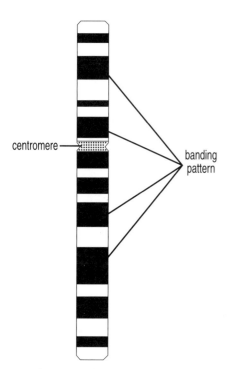

centromere

banding pattern

Figure 5.1
Diagram of a chromosome viewed through a microscope.

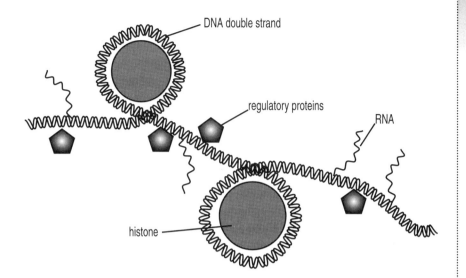

DNA double strand

regulatory proteins

RNA

histone

Figure 5.2
The arrangement of DNA, RNA,
and proteins within chromosomes.

▶ MATERIALS NEEDED

For each pair of students:

- 4 white or yellow pipe cleaners (also called chenille stems) (30-cm lengths)
- 4 cotton balls
- 3 felt-tip markers (green, blue, and brown)
- string or thread (approximately 16-cm lengths)
- cellophane tape
- 1 strip of Velcro (5-cm)
- 1 scissors
- 1 large plastic storage bag (1-gal)
- masking tape

▶ PROCEDURE

PART A: BUILDING THE MODEL

1. Shape two white or yellow pipe cleaners to represent a double helix (double strand) of a DNA molecule. Begin by twisting the pipe cleaners together at one end to attach them and then twisting them around your index finger to make one turn of the helix. Move your finger down the length of the pipe cleaner, twisting as you go (see Figure 5.3). When you have used up the length of the pipe cleaner, you should have a double helix with at least 5 turns.

Figure 5.3
Creating a model of DNA.

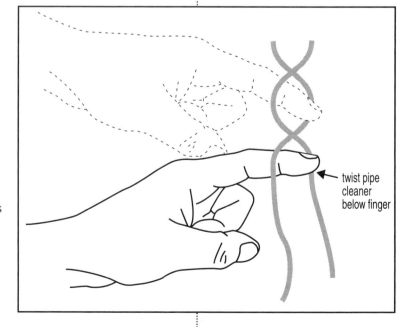

twist pipe cleaner below finger

2. With different colored felt-tip markers and Figure 5.4 as a guide, mark the location of genes on your chromosome.

3. Place 3 or 4 cotton balls, which represent histones, along the DNA helix by twisting the whole helix around the histone, as shown in Figure 5.2.

4. Cut 2.0-cm lengths of string or thread. Tape these threads, which represent RNA, onto your chromosome in any pattern you wish. (Review the Introduction and Figure 5.2.)

CAUTION: *Scissors are sharp; handle with care.*

PART B: REPLICATION

One of the major characteristics of life is the ability to reproduce or replicate. For a cell, this begins with the ability to replicate chromosomes faithfully.

NOTE:
- Remember that the discovery of the double helical nature of DNA explained how it might replicate itself.
- Remember what the end result of replication must be.

5. Remove the histones and RNA from your model. (A simplified model is easier to work with.) Take two more pipe cleaners of the same color as your model and model chromosome replication.

6. As you work through the model, record in your notebook what you are doing at each stage and how that represents the events that are happening in the cell as the chromosome is replicating.

7. Draw a diagram of your replicated chromosome in your notebook and describe the characteristics of this final product.

8. The two DNA strands of a single chromosome are attached at a region called a *centromere* which is located at different places on different chromosomes. Each double-stranded DNA is called a *chromatid*. Wrap the Velcro strip around each chromatid so that the hook portion wraps around one chromatid and the latch portion wraps around the other chromatid. This represents a centromere. Attach the chromatids.

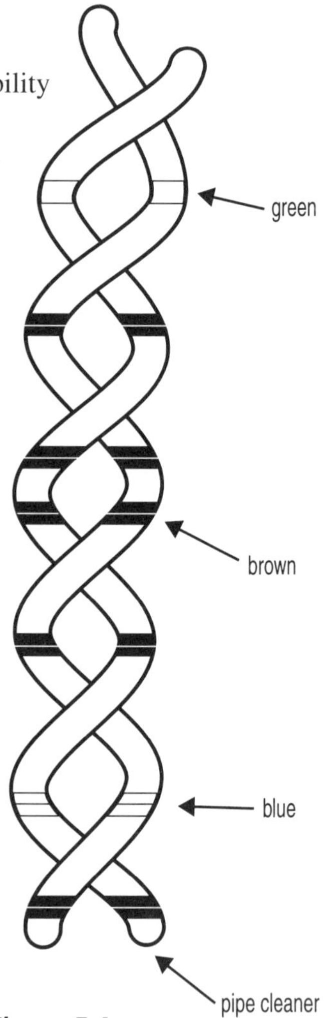

9. **STOP & THINK** How many centromeres and chromatids are represented on your chromosome? Label them on your diagram.

10. Place a piece of masking tape on a plastic bag. With a felt-tip marker, label the bag with yours and your partner's names. Place your model in the plastic bag for use in Learning Experience 6.

Figure 5.4
Placement of genes on chromosome.

green

brown

blue

pipe cleaner

CHARACTERIZING CHROMOSOMES

INTRODUCTION By the 1920s the concept of the chromosome was well accepted, but direct examination of chromosomes was not easy. The available dyes and stains did not reveal much detail; even locating and counting the chromosomes in a cell was chancy at best. The number of chromosomes in human cells was not certain.

In 1956, two geneticists, Joe Hin Tijo and Albert Levan, developed a simple and reliable procedure allowing direct visualization of an organism's chromosomes. It was known that in cells in the quiescent or nondividing state, chromosomes are spread out and diffuse. In fact, chromosomes seem to disappear in nondividing cells. During cell division, however, they form discrete structures within the cell. Tijo and Levan took advantage of this understanding and created a method for "catching chromosomes" when the DNA was tightly coiled and the chromosomes were most organized or condensed. They treated human cells with a drug to stop cell division at that point. Using pressure to squash the cells, the researchers spread them across a glass microscope slide and stained them to make the chromosomes more visible. They then photographed the chromosomes through the microscope. The final step was to clip each chromosomal image from the photograph, sort the images according to length, pair any matching sets of chromosomes, and paste them into a composite known as a *karyotype*.

Using this approach, Tijo and Levan were able to settle once and for all the matter of how many chromosomes were in a human cell—46. In addition, they confirmed in humans an observation made by Walter Sutton in grasshoppers in 1902, that in most cells, chromosomes come in matched pairs. That is, humans have 23 pairs of distinctly different chromosomes.

Investigations into a wide variety of organisms have found that most (but not all) organisms are diploid; that is, most of their cells carry two copies of each chromosome. While the number of chromosomes in different organisms varies widely (see Table 5.4), nearly every cell in a specific type of organism has the same set of chromosomes as every other cell.

In this activity, you will identify all the important features in a human karyotype.

Table 5.4

NUMBER OF CHROMOSOMES IN VARIOUS SPECIES	
SPECIES	**NUMBER**
Alligator	32
Ameba	50
Carrot	18
Chicken	78
Chimpanzee	48
Corn	20
Earthworm	36
Fruit fly	8
Garden pea	14
Goldfish	94
Grasshopper	24
Horse	64
Human	46
Sand dollar	52

▶ MATERIALS NEEDED

For each student:
- 1 sheet of plain white paper

For each pair of students:
- 1 "Human Karyotype" sheet
- 2 scissors
- 1 glue stick or cellophane tape
- 1 envelope

▶ PROCEDURE

1. You will be given a sheet containing one example of a human karyotype. Carefully examine the karyotype. Note the banding patterns and any other distinctive features that you can identify on the chromosomes.

2. Cut out each chromosome and match the pairs based on their identifying features.

3. Glue or tape the pairs on a new sheet of paper in rows, starting with the longest chromosome pair, and arrange them by size to the shortest. Number your chromosome pairs, starting with 1 for the longest pair.

▶ ANALYSIS

Write responses to the following in your notebook.

1. Describe several features that all the chromosomes in the karyotype have in common.

2. Describe the features of chromosomes that enabled you to identify the matching pairs.

3. Did every chromosome have a match? If not, why do you think this is?

4. This karyotype was made on a 14-year-old person. Will it look any different when this person is 60 years old? Explain your answer.

5. Based on your understanding of chromosomes and sickle cell, how many copies of the gene for hemoglobin does Denzel have? Explain your response.

READING THE FUTURE? ANALYZING CHROMOSOMES AND EXPLAINING THE CONSEQUENCES

INTRODUCTION   Is knowing what the future holds for you or your future children a good thing? What would you like to know about your future if you could know? Some people would like to know who their friends and partners might be, what kind of jobs they will hold, and whether they will have money or not. Others would like to know whether they will have good health.

Although there is no way to predict one's love life or financial status, chromosome and DNA analysis have opened up the possibilities of predicting, to some degree, a person's likely future health problems. A genetic counselor is a person who analyzes this information, helps individuals understand the science behind the analysis, and explains the complex implications of the results, including possible choices and outcomes.

Information about an individual's biological prospects is encoded in his or her DNA. An individual whose family has a history of a certain disorder might wish to know whether the information for this disorder is encoded in his or her genetic makeup. For example, in Learning Experience 2, Denzel Jones wanted to know whether he carried the sickling hemoglobin. One way to check for this is to do what Denzel did; have a test which checks for the variant protein. Another option open to Denzel was to have his DNA analyzed. In this case the blood taken from Denzel would have been used to isolate his DNA and to examine it for the single base change which results in the sickling hemoglobin.

Using a karyotype is another way to determine whether a person has a predisposition toward a genetic disease. In his investigations of fruit flies, T. H. Morgan observed that certain changes in the characteristics of organisms correlated with specific changes in chromosome structure or number. Today many genetic diseases have been correlated with changes in chromosomes which can be identified through karyotyping. Such changes may include a deletion of part of the chromosome, the addition of an extra chromosome (a *trisomy*), or a part of one chromosome may be exchanged with or added to another chromosome (a *translocation*).

Individuals may elect to have genetic testing done if there is a history of genetic abnormalities in their family. Since all cells in an organism (except red blood cells which lack nuclei) have the same set of chromosomes, any cell sample can be used for a karyotype. Normally white blood cells are used since they are easy to collect and good karyotypes can be obtained from them.

Expectant parents may wish to acquire information about their future child, especially if they have reason to believe (from family history or the birth of other children with inherited disorders) that the child might someday develop a disorder. In this case, samples are taken from the area of the developing fetus, either in the fluid surrounding the fetus (the amniotic fluid) or from one of the membranes (chorion) surrounding the fetus. These samples contain fetal cells which can be used to create a karyotype of the developing fetus. The karyotype can be examined for chromosomal aberrations that might result in disorders in physical and mental characteristics of the child. The fetal cells can also be tested for disorders such as sickle cell, cystic fibrosis, Huntington's chorea, and others that result from mutations in the DNA.

The role of the genetic counselor is to present the results of genetic analysis to individuals who have undergone testing and to provide information in careful clear, nonjudgmental and informative terms so that individuals who are confronted with possible indicators of their own or their baby's future health can make well-informed personal decisions.

In this activity, you will assume the role of a genetic counselor. The expectant parents have had prenatal testing done on their developing fetus because the mother is over age 40, and they know that the probability of chromosomal aberrations increases with increased age. The results have come in and your job is to determine, based on the karyotype, whether there is a problem. Normally a genetic counselor would give the parents the information in person, but for the sake of this activity, you will write them a letter.

▶ MATERIALS NEEDED

For each pair of students:
- 1 "Karyotype Placement Grid"
- 1 chromosome smear
- 2 scissors
- 1 glue stick or cellophane tape

▶ PROCEDURE

1. Obtain from your teacher the "photograph" of a chromosome smear and a copy of the "Karyotype Placement Grid."

2. Cut out the chromosomes from the chromosome smear and arrange them in pairs to create a karyotype on the "Karyotype Placement Grid." Use the normal karyotype you constructed as a guide to matching and arranging your chromosome pairs according to their banding patterns and sizes.

3. Write the number of your chromosome smear on the top of your "Karyotype Placement Grid." Glue or tape your chromosomes to the appropriate place on the sheet.

4. You may wish to determine the sex of the fetus (but remember, not all parents want to know ahead of time whether they will have a boy or a girl).

5. Your teacher will hand you a sheet entitled "Information on Chromosome Disorders." Determine whether the karyotype is normal or has any errors. If an error is apparent, identify the problem.

6. Discuss with your partner the presentation you will make to the class on your karyotype, the disorder, and its possible consequences.

7. Write a letter to the parents of the fetus that includes the following:
   – a basic introduction to chromosomes describing what they are
   – what information the DNA in the chromosome carries
   – how this information is transferred into protein
   – why functional proteins are important and what role they play for living things
   – why an error in the DNA or chromosome number or structure can result in variants in traits which can cause problems to the organism
   – how the karyotype was done
   – what information the karyotype can give
   – what the results of their fetus' karyotype have shown
   – if the results show a problem, how this problem might manifest itself in a child carrying this genetic makeup

   Conclude your letter by inviting the parents to come see you in order to discuss what alternatives they have and personal decisions they might make in light of the findings of the results. Remember that your letter must be clear, explain things simply yet completely, and be sensitive to the parents.

EXTENDING IDEAS

▶ Why *Y*? Why does the presence of the *Y* chromosome result in maleness in mammals? Scientists used to think that the male chromosome originally had many genes in common with the *X* chromosome (the other sex chromosome) but that in the course of evolution, the *Y* began to lose many of its genes, and in a sense, degenerate. Recent studies, however, present another, more "macho" possibility. Researchers have found evidence that the *Y* chromosome may actually be a refuge for genes that are beneficial to male fitness and fertility *(J. Travis, "The* Y *Copies Another Chromosome's Genes."* Science News, *November 16, 1996.)* Research the data for each of these theories, describe the evidence, and explain whether one theory is more convincing than the other and why you think this.

⬤ During the Olympic games in 1994, women competitors received Gender Verification cards when they passed a sex test. This test was required of all women athletes to authenticate the fact that they were actually women, not men masquerading as women. Karyotyping is done when there is a question of sexual identity. If any *Y* chromosomes are present, the athlete is declared ineligible. The rationale behind the testing is that maleness, as indicated by the presence of a *Y* chromosome, indicates the presence of androgens, hormones which cause increase in muscle mass and strength. Many individuals have protested that sex testing is unethical and may not be a true indicator of sex. Find out about sex testing and the controversy surrounding it. Decide whether you think it is a valid determination of sex and whether sex testing is ethical. Explain your decision.

⬛❶N THE JOB

GENETIC COUNSELOR   Jeanine and Richard Jensen have been married for 6 years. They had always planned on having a large family. Three years ago they were blessed with the birth of their daughter Michelle. Michelle seemed perfectly normal at birth, but after being home for a few weeks, Jeanine noticed that Michelle had stopped gaining weight and was coughing up a large amount of mucus. They brought Michelle in for a full physical, and after some tests, she was diagnosed with cystic fibrosis, a genetic disease whose symptoms include the malfunction of the pancreas and frequent lung infections. Michelle was very sick. Jeanine and Richard were heartbroken, but they were also relatively lucky. With daily medication and a special diet, Michelle has been able to survive for three years and there is hope for a long future, but the Jensens are concerned about the possibility that a second child could be born with the same condition. They love Michelle and they do not want to submit another child to the trials of cystic fibrosis even though the outlook for curing and for alleviating symptoms for the genetic disease seems brighter.

Jeanine and Richard made an appointment to see Susan West, a genetic counselor with the local medical center. Susan sat down with the Jensens and helped them consider the possible consequences of having another child. Since cystic fibrosis is a recessive gene, it means that both parents are carriers. The chances that one of their children would be born with the disease is 1 in 4. Susan helped the Jensens discuss their fears and their hopes. The presence of cystic fibrosis in a fetus can be detected by a genetic test during pregnancy. Should the Jensens decide to get pregnant, they had the option of this test. But a positive result could create great distress and bring about other decisions that would have to be made.

As a genetic counselor, Susan spends most of her time working with individuals and families, usually with a family history of disease. Some families contact her to find out if a particular disease is genetic and, if it is, then will they be able to have healthy children. Others see her after finding out that they are destined to develop a disease, such as Huntington's chorea, a disabling condition of the nervous system which does not usually develop until later in adulthood. Susan is there to help people make critical decisions in their lives by ensuring that people with little or no scientific background are able to make informed decisions as they are offered their first glimpse of DNA. For some people, there are no tests available at this time to detect whether they may one day have the possibility of contracting a particular disease. Yet, others have that option. Susan helps these people decide whether they want to know their fate and, thus, have the option of changing their lifestyle in ways to prevent it, or whether they would prefer not knowing, enjoying their life without the worry of what the future holds. Each day presents a challenge because no decision about one's fate is an easy one. Susan must stay objective, allowing patients to make their own informed decisions.

While in high school, Susan knew she wanted to work with people and she had a love of medicine. After going to college to major in biology, she was introduced to the science of genetics and found it very exciting. She soon learned that genetic counseling would combine her two interests. After graduating from a master's program in genetic counseling, Susan went on to work in a medical center known for its expertise in this field. Genetic counselors are presently in high demand. They can be found on the staffs of universities, in state offices, departments of social services and in bureaus of maternal and child health. They are employed in nursing homes, large hospitals, and many other institutions.

To revisit the Jensens, they decided to try to have another baby after discussing the information with Susan West, and while pregnant, made the decision to forego any further testing. Jeanine and Richard had a healthy baby boy. Susan was one of the first ones contacted with the good news!

LEGACY OF HEREDITY

PROLOGUE Contrary to movies, the media, and most popular opinion, the main purpose of sex (biologically speaking) is to produce offspring for the continuation of the species. To achieve that result, animals not only use sexual reproduction, but also have evolved elaborate behaviors such as courtship displays, fighting for mates, and sending candy and flowers on Valentine's Day. Why is all of that necessary? Many single-celled organisms can reproduce with incredible speed by simple cell division; a single bacterium can give rise to billions of cells in a few days, cells which are genetically identical to itself. As well as being speedy, this form of asexual reproduction is very efficient; even the most isolated cell can reproduce.

So why do organisms invest so much time, effort, and physical and even mental energy to reproduce sexually? Why has sex evolved? Several theories exist. The cells specialized for reproduction (sex cells or *gametes*) are the link to the future. Gametes pass information stored in their DNA from one generation to the next. As you saw in Learning Experience 4, mutations in the sequence of DNA sometimes occur. Gametes are endowed with mechanisms which enable them to repair many of these mutations more efficiently than other cells can. Reproducing through sex cells rather than by cell division is one way to ensure that random errors are passed on to offspring less frequently.

An additional explanation for the evolution of sexual reproduction is that this kind of reproduction allows for much greater variation in the characteristics of the offspring. Whereas the progeny of organisms that reproduce by asexual reproduction are almost identical to the parent (with a few mutations here and there), the progeny of sexually reproducing organisms show both variation from the parents and diversity among siblings.

In this learning experience, you will explore the processes by which information in chromosomes is passed from parent to progeny during sexual reproduction, and you will determine how this process can result in variation.

The Cells of Genetic Continuity

What, exactly, do we get from our parents, and how does it get from them to us? The legacy given to us by our parents is their chromosomes, which come to us packaged in their gametes. In order to understand this, we need to explore where gametes originate, how gametes develop, and what their purpose is.

During development of an organism, cells take on special functions in addition to their "housekeeping" functions of metabolism; muscle cells make proteins which enable the cells to expand and contract; cells in the brain make proteins which enable them to communicate by chemical and electrical interactions. Similarly, certain cells in the reproductive organs (ovaries or testes) become sex cells and develop the capacity to carry out the reproductive functions. Males develop sperm cells; females develop egg cells. Within these cells is all the information needed to create the next generation.

What do these sperm and egg cells look like, and how do they form? All human body cells except sex cells contain 46 chromosomes. *Meiosis* is the process that produces sex cells—a process that reduces the chromosome number in half.

The female reproductive cell (ovum or egg) develops by a process called *oogenesis*. At birth a human female contains about 400,000 primary oocytes in her ovary. These oocytes contain the same number of chromosomes as every other cell in the body—46 (the *diploid* or full chromosome number) or 23 pairs. During its development, the oocyte matures, undergoes division, and its chromosome number is reduced from 46 to 23, the *haploid* number. Four cells are generated from the primary oocyte; three small

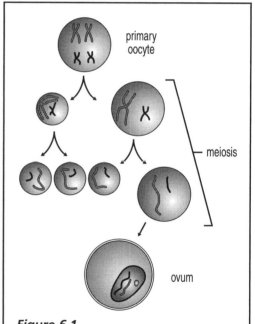

Figure 6.1
During oogenesis eggs mature in the ovary of the female. Primary oocytes, which contain 46 chromosomes, undergo meiosis to produce ova or mature eggs. (Only 4 out of 46 chromosomes are shown.)

cells that each contain very little cytoplasm and 23 chromosomes, and one larger cell that contains most of the cytoplasm of the original oocyte and 23 chromosomes. (See Figure 6.1.) This larger cell, the egg (ovum), contains one copy of each chromosome of the woman and can be fertilized by the sperm. At monthly intervals after puberty, one egg is released from the ovary (ovulation) and makes its way down the Fallopian tube to the uterus (see Figure 6.2). If the egg is fertilized, it will attach to the uterine wall and develop; otherwise it will be released in the monthly menses. During a woman's reproductive life, approximately 400 eggs will make this journey.

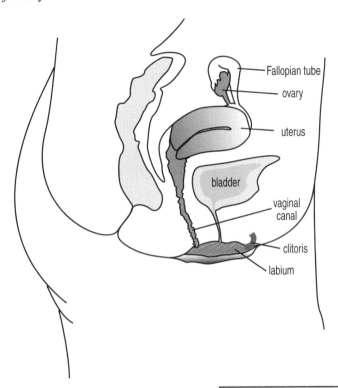

Figure 6.2
Once a month an egg makes its way from the ovary through the Fallopian tube to the uterus.

A similar process of chromosome reduction or meiosis occurs during sperm development or *spermatogenesis*. The primary spermatocyte also has 46 chromosomes. It develops in the sex organ of the male, the testis, and undergoes meiosis generating four cells, each of which contains 23 chromosomes. Further maturation results in four sperm cells with compact head-pieces and long, powerful flagella or tails which enable them to move. Unlike in the female where only one mature egg is produced for each oocyte, a primary spermatocyte will produce four mature sperm (see Figure 6.3). After maturation is complete,

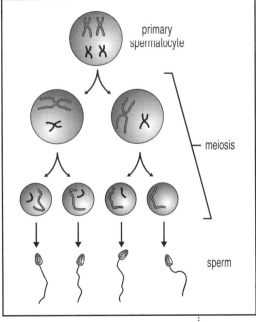

Figure 6.3
During spermatogenesis sperm mature in the testes of the male. (Only 4 out of 46 chromosomes are shown.)

sperm leave the testes and travel through a system of ducts or tubules which produce fluids that help move the sperm to the opening of the penis where they are released (see Figure 6.4). After puberty, males continually produce vast numbers of sperm—approximately 200 million per day.

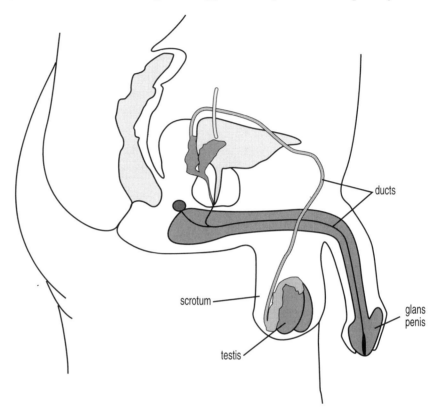

Figure 6.4
From the testes, sperm move through a system of ducts which produce fluid that helps move the sperm toward the opening of the penis for release.

▶ ANALYSIS

Write responses to the following in your notebook.

1. Compare and contrast (in chart form or a labeled diagram) oogenesis and spermatogenesis.

2. Do you think each of the four resulting gametes gets an identical set of 23 chromosomes? Why or why not?

3. What would be the result if meiosis did not occur?

4. Fertilization is the joining of the male and female gametes. Draw or describe the chromosomal composition that results from this union. Why is this result important?

Dance of the Chromosomes

INTRODUCTION The process of meiosis has been referred to as "the dance of the chromosomes" because of the way the chromosomes move and sort. They have also been compared to marionettes dancing and moving on strings. In this activity, you will build two additional sets of chromosome models to complete the chromosomal makeup of a mythical mosquito. A karyotype of this mosquito has revealed that it has three pairs of chromosomes (which is the normal number for mosquitoes). After building your models, you will demonstrate the process of meiosis and determine which genes will be in the mosquito gametes. The genes which an organism has are called its *genotype*. You will determine the *phenotype* (that is, what an organism looks like as a result of the expression of the genes) of these offspring.

 MATERIALS NEEDED

For each group of four students:

- 2 models of chromosomes (one white and one yellow) with two chromatids, stored in large plastic bags (constructed in Learning Experience 5)
- 16 pipe cleaners (chenille stems)
  - 4 each yellow and white (22-cm)
  - 4 each yellow and white (15-cm)
- 3 felt-tip markers (red, black, and brown)
- 4 strips of Velcro (5-cm lengths)
- masking tape
- 1 coin
- 1 large plastic storage bag (1-gal)

PART A:

▶ **PROCEDURE**

1. Gather all the materials for your group. Your group will consist of one pair of students that has constructed a yellow chromosome and one pair of students that has constructed a white chromosome.

2. Place your group's yellow and white chromosomes on the table.

3. Construct four more replicated chromosomes, one of each color (yellow and white) and one of each length (22-cm and 15-cm).

NOTE: For the purposes of this activity you do not need to put on histones, other proteins, or the RNA, but be sure to bear in mind that these are always present.

As in the model you made in Learning Experience 5, these models consist of:

- two intertwined pipe cleaners which represent the DNA of the chromosome
- two colors (yellow and white) which represent the contributions of the two parents (white from the mother, yellow from the father)
- colored bands drawn on the DNA to represent specific genes (use Figure 6.5 as a guide to gene location)
- Velcro, which represents the centromere (2 loops = 1 centromere)

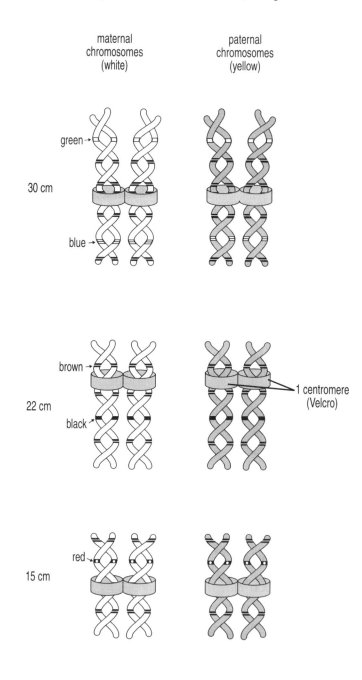

Figure 6.5
Each replicated chromosome model should have:

- two chromatids containing four DNA strands (two double helices)
- one centromere
- genes

4. Genetic studies of this mythical mosquito have identified four genes which are responsible for four of its traits. Use Figure 6.5 and mark the genes on each of your two new chromosomes.

PART B:

Your group now has six mosquito chromosome models representing three pairs of *homologous chromosomes* (chromosomes that have genes for the same trait at the same location)—three maternal and three paternal. Each chromosome pair contains one or two genes whose traits are known and several genes whose functions are not known:

- Red bands represent the genes that encode a protein responsible for antenna texture (either hairy, represented by the letter *H,* or smooth, represented by the letter *h*).
- Green bands represent the genes that encode a protein that determines wing shape (round is *R,* or square is *r*).
- Black bands represent the genes that encode a protein that determines wing length (long is *L,* or short is *l*).
- Blue bands represent the genes that encode a protein that determines abdomen color (black is *B,* or orange is *b).*
- The brown bands indicate other genes on the DNA, whose encoded functions are either unknown or not of interest in this study.

Table 6.6
Summary of information about the traits of the mosquitoes.

TRAIT	PHENOTYPE	SYMBOL FOR ALLELE	BAND COLOR	CHROMOSOME (cm)
antenna texture	hairy smooth	*H* *h*	red	15
wing shape	round square	*R* *r*	green	30
wing length	long short	*L* *l*	black	22
abdomen color	black orange	*B* *b*	blue	30
unknown	not available	not available	brown	on all chromosomes

The labeling of the genes in this activity reflects the convention used by geneticists. The capital (upper case) letter represents the dominant trait; and the small (lower case) letter, the recessive trait. (For example, in peas the gene for the trait of pea shape is either *R* for round, meaning it encodes a functional SBEI, or *r* for wrinkled, which means the gene encodes a faulty or non-functional SBEI). Since an *allele* is a different form or variant of a gene for a specific trait, for example, the gene for shape has two alleles: the allele that results in wrinkled shape and the allele that results in round shape.

In this activity, each gene has two possible alleles. Which allele is present on the chromosome will be determined by coin flipping. To assign the allele of your four genes, your group must make a total of eight coin flips. Both chromatids on a chromosome will have an identical allele.

▶ PROCEDURE

1. Locate the red bands representing the allele *H* or *h* on your white 15-cm chromosomes.
  – Flip the coin once for the white 15-cm chromosome. If it lands "heads," write the letter *H* on two small pieces of masking tape and attach the tapes to the red bands on the chromatids of this chromosome. If it lands on tails, write the letter *h* on two pieces of masking tape and attach to the red bands on the two chromatids.
  – Flip the coin a second time for the yellow 15-cm chromosome, and label the alleles (*H* or *h*) as described above.

2. Repeat Procedure step 1 for determining the allele on each chromosome for the three remaining traits—wing shape (*R* or *r*), wing length (*L* or *l*) and abdomen color (*B* or *b*). Use Table 6.6 as a reference for gene location.

3. **STOP & THINK** Your group now has specific alleles for each of the four known genes on your six mosquito chromosomes. These collective alleles represent the genotypes of the two parent mosquitoes. An example of one possible result for one parent mosquito is shown in Figure 6.7. Record the genotype for your mosquito in your notebook.

4. **STOP & THINK** Record in your notebook all the parts of the six models with descriptions or labeled diagrams where appropriate. Include the following:
  – what each different color chromosome represents
  – why there are two chromatids in each of the chromosomes
  – how you would explain what *R* and *r* represent
  – what the relationship is between *H* and hairy antennae
  – why the alleles for each gene in each chromatid are identical within the same chromosome
  – why the alleles in the homologous chromosomes might be different
  – what the difference is between an organism's genotype and its phenotype

5. At the end of the session, store your models in the bags for use in the next session.

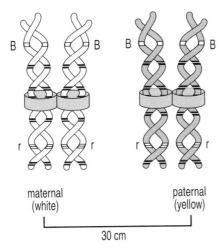

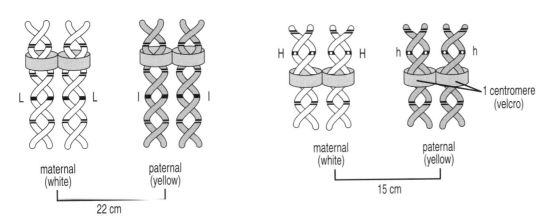

Figure 6.7
One possible mosquito genotype resulting from the coin flips. The genotype for this mosquito would be *BB, rr, Ll, Hh*.

SIMULATING THE STAGES OF MEIOSIS

INTRODUCTION In this activity, you will model the events that occur in sex cells as your mosquito becomes sexually mature. The chromosomes you have built are found in the primary oocytes or primary spermatocytes. Recall from your reading that when an organism matures sexually, its gametes or sex cells develop from their precursor cells—oocytes or spermatocytes—into eggs or sperm through the process of meiosis. In the activity, you will examine how the chromosomes move and sort themselves. Table 6.8 describes and illustrates each stage of meiosis. Use the table as a reference to help you complete the Procedure in which you will model the process of meiosis.

Table 6.8

▶ MEIOSIS I: STAGES OF MEIOSIS

A	The oocytes or spermatocytes (precursors to the gametes) have undergone mitosis (cell division) and are ready to mature. In the mosquito cell, there are six chromosomes (or three pairs). At this point, the DNA within the chromosomes has replicated (or doubled). There are six chromosomes, each having two double strands of DNA (called chromatids) joined together by a centromere. The nuclear membrane begins to break down. This stage is called Interphase I	
B	The homologous pairs of chromosomes line up and spindle fibers (protein structures in the cell which serve to guide the chromosomes around the cell) start to form. This stage is called Prophase I.	
C	The pairs of chromosomes line up along the equatorial plane of the cell. This stage is called Metaphase I. Each different length pair consists of two homologous chromosomes (four chromatids).	
D	During Anaphase I, the chromosomes move away from the center of the cell.	
E	When chromosomes reach the ends of the cell, the cell begins to divide to form two daughter cells, each with one copy of each chromosome. This stage is called Telophase I.	
F	Cell division is complete, the nuclear membrane reforms and the chromosomes become less discrete in the cell. This stage is called Interphase II.	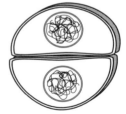

G	In each of the two cells formed at the end of Meiosis I, chromosomes condense into discrete units and spindle fibers form. This is Prophase II.	
H	During Metaphase II, chromosomes line up along the center of the cell.	
I	The attachment between the two chromatids of each chromosome breaks. Chromatids move away from each other along the spindle. This phase is called Anaphase II.	
J	In Telophase II, cell division begins, forming two new daughter cells—gametes.	
K	Cell division (cytokinesis) is complete. The nuclear membrane reforms. Proteins associate with the chromatids to form new chromosomes. Chromosomes become less discrete.	

▶ MATERIALS NEEDED

For each group of four students:

- 6 complete chromosome models from "Dance of the Chromosomes"
- 1 piece of yarn (6–8 m)
- 1 scissors
- 1 measuring tape or meter stick

▶ PROCEDURE

1. Cut three pieces of yarn: two 1 m long and one 2 m long. Use the larger piece to form a circle on a table top. This represents the cell membrane of your sex cell. Use one smaller piece to form a circle within the large circle. This represents the nuclear membrane.

2. Be sure the chromatids in each chromosome are joined at the centromere with Velcro, and then place all of your chromosomes anywhere within the inner circle. This represents the beginning of meiosis. At this stage, the chromosomes have replicated and formed chromatids (row A of Table 6.8).

3. Pair each chromosome with its homologous chromosome (same size, same genes). Remove the inner circle. The nuclear membrane breaks down at this point in meiosis (row B).

4. Place your chromosomes in a straight line within the cell (row C).

5. Separate your homologous chromosome pairs into two rows, each row consisting of one each of the different chromosomes. Move the rows apart to opposite sides of the cell (row D).

6. At this point, a new cell membrane is formed (row E) and two daughter cells result. Model this by cutting the outer yarn (cell membrane) in half and closing each half around one set of three chromosomes. (You may wish to attach additional pieces of yarn to enlarge each cell membrane—representing growth.) Place an inner circle of yarn (nuclear membrane) around each of the two newly formed nuclei in the daughter cells. During this stage, the chromosomes become less visible (row F).

NOTE: If you have to put your models away for any reason, be sure to note the arrangement of your daughter cells and chromosomes for modeling Meiosis II later.

▶ ANALYSIS

Write responses to the following in your notebook.

1. Describe in your own words the events of this first meiotic division. Indicate the lineage (maternal or paternal) of each chromosome in each cell.

2. What other chromosomal patterns might have resulted in the daughter cells? (You may wish to move your model pieces around to help illustrate the possibilities.)

3. How many chromosomes does each daughter cell now have? What helped you determine that number?

4. How many alleles for each trait does each daughter cell have?

PART B: MEIOSIS II

▶ PROCEDURE

1. At the end of Meiosis I, two complete daughter cells have formed (Row F of Table 6.8). At this stage, chromosomes condense and become visible again (row G). Remove the inner circle of yarn. (The nuclear membrane breaks down.) Line up the three chromosomes vertically in the middle of each cell within your two daughter cells (row H).

2. Within each cell, separate the chromatids by detaching the Velcro. Move the separated chromatids away from the middle, toward opposite sides of the cell (row I).

3. Use your pieces of yarn to represent cell membranes that form as each cell divides. A total of four gametes have formed from each of the starting sex cells, each containing one copy of each chromatid which now looks like a chromosome. The nuclear membrane reforms. Model this using more yarn (rows J and K).

4. Store your models in a bag for use, once more, in the next session.

▶ ANALYSIS

Write responses to the following in your notebook.

1. Describe in your own words what has taken place in Meiosis II. Indicate which chromosome came from which parent in each of the resulting cells.

2. What is the difference between the starting cell (the primary spermatocyte or oocyte) and the cells at the end of Meiosis I? Between the cells at the end of Meiosis I and the cells at the end of Meiosis II?

3. What other maternal and paternal chromosomal combinations might have resulted?

4. Describe why you think meiosis is a significant process in terms of:
   – the passing on of information from generation to generation
   – variation of traits among siblings

5. Write the genotype of each of the four gametes in your group's model. Use the symbols for the alleles for each chromosome.

THE MATING GAME

INTRODUCTION When fertilization occurs, gametes fuse and the genes from two individuals join. It is this new combination of genes that results in the formation of a unique individual—be it an insect, animal, or plant.

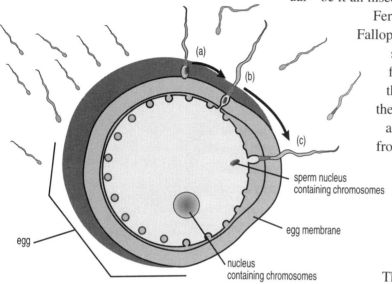

Fertilization in humans occurs in the Fallopian tube. Males release millions of sperm into the vaginal canal of the female. These sperm squirm and thrash their way through the uterus and enter the Fallopian tube where they encounter a mature egg that has been released from the ovary. Sperm compete to donate DNA to the female egg. Of all the millions of sperm that surround the egg—which is approximately 75,000 times larger than a sperm cell—only one may bind, penetrate, and insert its DNA into the egg (Figure 6.9).

The fertilized egg (or zygote) continues its journey through the Fallopian tube to the uterus, eventually attaching to the uterine wall where it continues to develop into a multicellular organism.

The joining of the chromosomes of the egg and sperm results in restoring the full complement of DNA for the zygote. Having the complete number of chromosomes, the zygote can begin development.

In the following activity, you will simulate fertilization by combining the genetic material from the male and female gametes of a mythical mosquito. You will then determine the genotypes and phenotypes of the offspring.

Figure 6.9
Fertilization of an egg. A single sperm (a) binds to the egg, (b) penetrates, and (c) injects its DNA.

▶ **MATERIALS NEEDED**

For each group of eight students:
- 2 large plastic bags containing chromosome models from the meiosis activity

▶ **PROCEDURE**

NOTE: It does not matter which color you choose. You do not need the sister chromatids.

1. In your group of four, place your group's chromosomes on the table.
2. Choose one chromosome of each length (15-cm, 22-cm, 30-cm). Return the remaining chromosomes to the plastic bag.

3. Your teacher will designate half the groups as male and half the groups as female. Each group designated as male should join a group designated as female.

4. Copy Table 6.10 into your notebook and write in the genotypes of the "sperm" and the "egg" of the fertilized mosquito cell.

Table 6.10

	ANTENNA TEXTURE *H* or *h*	WING SHAPE *R* or *r*	WING LENGTH *L* or *l*	ABDOMEN COLOR *B* or *b*
genotype of egg				
genotype of sperm				

5. Use your chromosome models to model fertilization between the two gametes. Pair up the homologous chromosomes.

6. Record the genotype of the resulting offspring.

7. **STOP & THINK** Based on the genotype, what is the offspring's phenotype? Record the characteristic for all four traits. (You may wish to refer to Table 6.6.)

MISTAKES OF MEIOSIS

INTRODUCTION The human karyotype on the next page (see Figure 6.11) shows three chromosome 21s, a trisomy. Using your knowledge of how chromosomes move and sort during meiosis and of fertilization, write two or three paragraphs in your notebook which explain, in detail, how this individual might have received three of chromosome 21. Include diagrams which show:

- how chromosome 21 moved and sorted to result in the trisomy

- which gametes were involved in the fertilization

- the other gametes produced at the same time.

Explain whether other gametes are affected. And if so, how?

Figure 6.11
A trisomy 21 karyotype.

READING

CLONING OF THE LAMB: SILENCE OF THE MAN?

In February of 1997, the scientific world was taken by surprise by the announcement by Dr. Ian Wilmut of the Roslin Institute outside of Edinburgh, Scotland that his group had cloned a sheep from the cells of the mammary gland of another sheep. The cloned sheep, an identical copy of the sheep from which the cells were taken, was born in July 1996 and was named "Dolly" (after Dolly Parton).

The techniques involved were remarkably simple considering the momentous achievement. The scientists took cells from the udder of a six-year-old Finn Dorset ewe and placed them in an extract low in nutrients. By starving the cells, scientists were able to switch off the genes that were being expressed. Then they took an oocyte from the ovary of a Blackface ewe and removed its nucleus, leaving the cytoplasmic material from the egg, with all its proteins required to produce an embryo, intact.

The mammary cell was placed next to the enucleated oocyte in a dish and the cells were stimulated with an electric pulse. This caused the two cells to fuse together and the nucleus of the mammary cell, with its DNA entered the oocyte. A second pulse caused the cell to start to divide and begin embryonic development. Six days later, the ball of embryonic cells was implanted into the uterus of a Blackface ewe where it continued to develop. The birth of Dolly occurred on the 277th try, of which only 29 had developed enough to reach the implantation stage.

The implications of this landmark mammalian cloning experiment—previous experiments in frogs had reached the tadpole stage, but they had all died before becoming adults—was of great interest and concern to the general public.

▶ ANALYSIS

Write responses to the following in your notebook.

1. Is Dolly a Finn Dorset or a Blackface sheep? Explain your response.

2. At the announcement of this feat, was Dolly seven months old or over six years old? Explain your reasoning.

3. Do you think this experiment could have been successfully done in the other direction, that is, placing the oocyte nucleus into an enucleated mammary cell? Explain your response.

4. What concerns do you think have surfaced at the news of this cloning? Include what concerns you may have.

5. What benefits do you think might result from this type of experiment?

6. Describe the implications that cloning may have for the role of the male of the species. What implications might this have for biological variation?

EXTENDING IDEAS

In March of 1997, there were reports that a group of scientists in Oregon had produced monkeys from cloned embryos. This is the first time a species so closely related to humans has been cloned. The scientists used a technique similar to the one that the Scottish scientists used to clone Dolly, the sheep. The two monkeys, born in August 1996, were cloned from cells taken from embryos, not an adult as in the case of Dolly. Therefore, the cloned monkeys are not genetically identical to any adult monkey.

Research both experiments, examining the science and the ethical issues, and write a position statement on the moral and ethical considerations of these experiments. If you feel that there are ramifications evident in one, but not the other, state your reasons for the difference.

ON THE JOB

RECREATIONAL THERAPIST   Manuel's two o'clock class will start shortly. On this particular day, Manuel will be leading his class in learning the latest new line dance. He's had a busy day already, working with numerous young adults at a school for the mentally disabled. This particular class includes 10-year-old Lucinda. Lucinda was born with Down syndrome. Like others in the class, Lucinda feels clumsy and has difficulty performing certain tasks. She was very shy and withdrawn when she met Manuel a few months ago.

Manuel's job as a recreational therapist is to help children like Lucinda have fun while learning coordination, increasing their self-esteem, and socializing effectively through physical activities. He does this by engaging students in music and dance, arts and crafts, cooking, and playing different sports. Even though much of his time is spent in a class situation, he also spends time with each student separately. While some children will grow up to be semi-independent adults, others will remain dependent upon others for care. Manuel uses information from medical records, medical staff, family, and the patients themselves to help individual students, working at their personal skill levels. This may include something as basic as tossing a ball back and forth or as complicated as playing volleyball. He helps train a number of his students for the annual Special Olympics and has seen many of them win medals.

While in college to complete a bachelor's degree in recreational therapy, Manuel learned that positions were available in hospitals, adult day care centers, nursing homes, and retirement facilities, as well as residential facilities for the disabled. He could have found work in a nursing home by getting an associate's degree, but Manuel always enjoyed the thought of working with children in a residential setting, which requires more education.

The class has ended. Lucinda loved the music and picked up the moves right away. Manuel chose her as his special helper to aid others learn the dance. He loves his job, feeling he makes a difference in people's lives while bringing smiles to their faces and giving them a sense of accomplishment.

IN THE ABBEY GARDEN

PROLOGUE   **A**n old adage says that in life we have to "play the cards we've been dealt." If we rephrase that adage from a molecular perspective, we could say that in life we play the DNA we've been dealt. In many ways, the cards (or DNA) you get in life depend on your starting deck (the genes from your parents) and the way the cards have been shuffled (in meiosis). The deal may be random and the cards seem to come up by chance, but there is a certain predictability, as any gambler or magician knows!

Imagine you were to arrange a deck of cards in the following specific order—the aces of hearts, clubs, diamonds and spades; then the twos of hearts, clubs, diamonds, and spades; and so forth through the kings. If you then dealt the cards out in four piles, with the ace of hearts in the first pile, the ace of clubs in the second pile and so forth in order, you would have 100% certainty (or probability) that each pile contains the ace through the king of a single suit.

But, if you shuffled the original deck before making the piles, the four piles would look quite different from each other. The order and suits in each pile would be random, with no two stacks alike, although the probability of a single card ending up in any one pile could be mathematically calculated.

The assortment and recombining of chromosomes during meiosis and fertilization are very similar to the shuffling and distributing of cards from a deck. In this learning experience, you will explore how the distribution of chromosomes to offspring is a random process, but such distribution can be predicted by the laws of probability. You will examine the experiments of Gregor Mendel, whose investigations into the inheritance patterns of peas led to the establishment of several fundamental principles of inheritance and to an understanding of the variation in the traits that are observed between parents and offspring, and among siblings.

THE FLOWERS THAT BLOOM IN THE SPRING

INTRODUCTION In every flowering plant, whether it is an apple tree or a dandelion, the flower is the reproductive organ that allows for the continuation of the species. Despite the extraordinary diversity in their shape, coloration, and size, flowers, like the plants that bear them, demonstrate simplicity and similarity in structure and function.

Within flowers, gametes are produced, fertilization occurs, and the seeds develop. Some flowers, such as a lily, contain both male and female parts. The female structure is in the center of the flower and consists of a vase-shaped ovary, where the eggs are produced, and a stigma which connects to the ovary through a tubular style. The more numerous male parts of the flower usually surround the female parts and include the anther which produces the pollen (or sperm) and the filament which supports the anther. Surrounding these male and female parts are the petals and sepals, which are not only attractive to humans but also attract insects, birds, and other animals that play an important role in plant fertilization (see Figure 7.1).

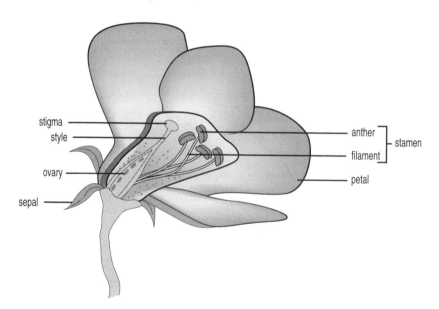

Figure 7.1
Reproductive structures in a flower.

In this investigation, you will conduct a thorough study of a flower. You will need to make detailed observations, dissect one flower, and locate and identify the structures involved in sexual reproduction in plants.

▶ MATERIALS NEEDED

For each pair of students:

- 2 fresh flowers (from the same type of plant)
- 1 scalpel or single-edged razor blade
- 1 dissecting needle
- 1 hand lens
- paper towel or shallow tray
- biology textbook (optional)

▶ PROCEDURE

1. **STOP & THINK** Before you begin the dissection, look closely at one of the flowers. Draw and describe your observations of the flower in your notebook.

2. Set aside one flower, intact, for comparison later.

3. With the scalpel and dissecting needle, begin dissecting one flower. Carefully remove and place each component part on a paper towel or a tray. As you work, identify each structure. Note how the location and structure of the parts relate to their functions.

4. When you have completed the dissection, describe each of the parts separately with a labeled diagram.

CAUTION: Use caution when handling sharp objects.

▶ ANALYSIS

Write responses to the following in your notebook.

1. Create a concept map which demonstrates the relationships among the structures of a flower, their functions, and the process of fertilization.

2. In nature, the pollen from male plants is often carried to female plants by insects which have been attracted to the nectar found in the brightly colored flowers. This method of mating results in random mixing of variants of traits of different plants. Plant growers often wish to breed plants in order to produce progeny with very specific characteristics. To do this, the mating or crossing of the parent plants must be done in a controlled manner, with careful attention paid to which plants cross with which. Using your understanding of plant reproduction, design a method which a plant breeder might use to create a new variety of plant. Describe each step of the process and the events which occur in the plant after pollination.

THE PARSON AND HIS PEAS

Throughout history, farmers and gardeners have been interested in how traits are passed from generation to generation in their livestock and crops. By studying patterns of inheritance and by carefully choosing the parents to be used in breeding, they could produce plants and animals with specific characteristics. It was amid this interest in developing new varieties of plants and animals that a modern day understanding of heredity began to emerge.

Gregor Mendel was not a scientist; he was a monk who took a great interest in the plant varieties growing in the garden at the abbey where he lived. Unlike some gardeners of the time who bred plants in order to create new combinations of flower color, leaf shape, and hardiness, Mendel was interested in breeding as a way to investigate how traits were passed from generation to generation. Mendel took a special interest in garden peas. With the help of the other gardeners at the abbey, Mendel isolated several strains of plants with distinct traits: one produced only tall plants, another only short plants; one produced only purple flowers, another only white flowers.

Mendel chose to follow seven traits in peas, each of which showed only two variations. This made it easy for him to track the phenotypes. The phenotypes themselves—round or wrinkled seeds, yellow or green seeds—were easy to observe with the naked eye. Figure 7.2 shows the traits Mendel selected.

Other breeders had already observed that for some traits, one form seemed to show up much more frequently than the other. Mendel's experiment involved crossing plants with different variants of the same trait, observing and counting which form showed up in the offspring, and analyzing the inheritance of the variant.

If Mendel wanted to breed a pea plant having round peas with another pea plant having round peas, how could he do it? Mendel used his understanding of plant reproduction to design an experimental approach. To follow patterns of inheritance, he needed to control which plants crossed with which to produce offspring.

Mendel knew that peas are self-pollinating (also called self-fertilizing). Because pea flowers develop male and female parts simultane-

Figure 7.2
The seven traits Gregor Mendel chose for his experiments, and the two variations of each trait.

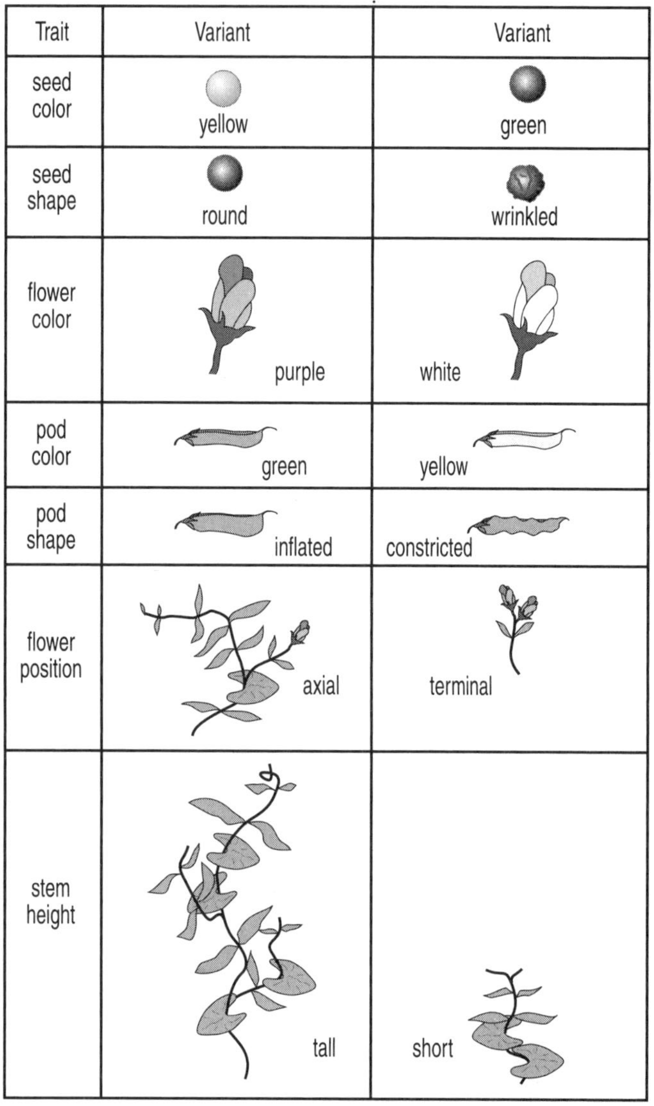

Trait	Variant	Variant
seed color	yellow	green
seed shape	round	wrinkled
flower color	purple	white
pod color	green	yellow
pod shape	inflated	constricted
flower position	axial	terminal
stem height	tall	short

ously, the plant may fertilize itself, depositing the pollen from its anther onto its own stigma. In different plant varieties in which the male and female parts mature at different times, the flower does not pollinate itself. In these cases, pollen is carried to the stigmas of other plants of the same species primarily by insects, but also by birds, other animals, or wind. As the insect feeds, the pollen from the anthers sticks to its body. As the insect moves from flower to flower, it leaves a "calling card" on each, pollen which attaches to the stigma, initiating fertilization. The process by which pollen from one plant fertilizes the flowers of another plant is called cross-pollination or cross-fertilization.

The method that Mendel used to breed his pea plants involved several steps: He opened the pea flowers and removed the anthers. He then dusted the stigmas with pollen from another plant (see Figure 7.3).

In this way, he could carry out controlled pollination from one plant to another. After each cross, Mendel collected seeds and recorded the traits of the seeds that resulted from each cross. He then planted these seeds in his garden; when these plants matured, he determined the traits of the plants. These seeds and plants were called the F_1 (first filial) generation. He then carried out controlled crossings of the F_1 plants and again noted which seeds came from each cross. The resulting plants were called the F_2 generation. Using this approach, Mendel collected and analyze data on thousands of offspring through many generations.

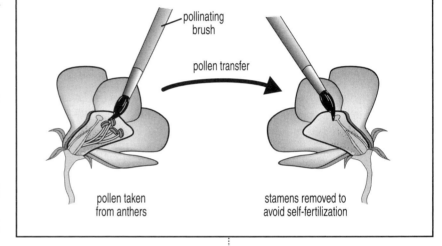

Figure 7.3
Mendel carried out cross-breeding by brushing pollen from one plant onto the stigma of another.

MY PARENTS ARE ROUND AND I'M WRINKLED; WHAT GIVES?

Prior to beginning his experiment, Mendel had observed many generations of plants to ensure that they always produced the same phenotype for a specific trait—for example, plants with round peas always gave rise to new plants with round peas.

THE FIRST GENERATION
To breed a first generation of pea offspring, Mendel took great care to cross-fertilize plants having differing variants of traits. He began by

crossing plants which only produced wrinkled peas with plants that only produced round peas (the symbol for crossing is "x," and so this experiment can also be represented as "round x wrinkled"). Mendel took pollen from plants that produced wrinkled seeds to fertilize the flowers of a strain that produced round seeds. He also did the reverse: he used the pollen from plants that produced round seeds to fertilize the wrinkled strain. One of his first observations was that the results did not change depending upon which plant contributed the pollen and which contributed the egg. In the first generation of offspring (the F_1), he obtained the following data for the number of round seeds and the number of wrinkled seeds produced:

	round x wrinkled ("parents")	
	round seeds	wrinkled seeds
F_1 generation	7300	0

Mendel's published data did not indicate how many plants he crossed to produce all the seeds observed in the F_1 generation. What is known is that the flowers were carefully cross-fertilized, every pea pod was collected, and all the seeds individually sorted and counted as either wrinkled or round. Mendel carried out these crosses many times in order to obtain statistically significant data.

THE SECOND GENERATION
To examine the second generation, Mendel crossed plants from the F_1 generation. He then collected and analyzed the seeds produced by this cross. His data from counting the shape of 7324 F_2 generation seeds follows.

	round x round ("parents")	
	round seeds	wrinkled seeds
F_2 generation	5474	1850

One of his observations about seed shape, an observation that held true with all seven traits shown in Figure 7.2, was that one phenotype always appeared in the F_1 generation. He called this the *dominant* variant. The other variant seemed to disappear in the F_1 generation and to reappear in the F_2 generation; he called this the *recessive* variant. With only the evidence of his garden pea crosses, he established several important genetic principles, including:

- There are pairs of "factors" that control heredity (now called genes). In organisms that reproduce sexually, genes are inherited from each parent.

- In cases in which an organism possesses two forms of the gene for a single trait, one form of the gene may be dominant and the other may be recessive. This is known as the *principle of dominance.*

- Two forms of each gene are separated (or segregated) during the formation of reproductive cells. This is known as the *principle of segregation.*

- The genes for different traits may assort independently of one another. This is known as the *principle of independent assortment.*

Unfortunately, no one was ready to accept Mendel's work when it was completed in the 1860s. His paper, which detailed his results and conclusions, was generally ignored for more than 30 years after its publication. It is difficult for us to comprehend today just how incredible Mendel's achievement was. Mendel did his work prior to the knowledge or understanding of genes, chromosomes, DNA, and meiosis. Mendel's principles provide the basis for much of the current thinking in genetics.

▶ **ANALYSIS**

Write responses to the Analysis in your notebook.

1. Select one trait (other than seed shape) and construct a flow chart or diagram of the steps Mendel used to collect his data. In Table 7.4, the dominant variant is listed in bold print.

2. What is the genotype of the round seeds in the "parent" generation? In the F_1 generation? Of the wrinkled seeds in each generation?

3. Do you think the genotype(s) of the round seeds in the F_2 generation might be the same as in the F_1 generation? Explain your answer.

4. In the F_2 generation, 7324 seeds were collected: 1850 were wrinkled and 5474 were round. What is the approximate ratio of these numbers? Why do you think there are many more round seeds than wrinkled seeds?

5. Explain in a paragraph or diagram how an understanding of meiosis explains Mendel's principles of segregation and independent assortment.

Table 7.4

TRAIT	ALLELES
seed color	**yellow**, green
seed shape	**round**, wrinkled
flower color	**purple**, white
pod color	**green**, yellow
pod shape	**inflated**, constricted
flower position	**axial**, terminal
stem height	**tall**, dwarf

SORTING THE CROSSES

Are the patterns or ratios in inherited traits predictable or random? If segregation and assortment occur the way Mendel thought they did, then the possible gene combinations in the offspring that result from a cross can be predicted.

In the previous learning experience, you followed the segregation of genes during the formation of the reproductive cells (meiosis). Suppose the F_1 plants have one tall allele from one parent and one short allele from the other parent. The plant grows and forms flowers and the two alleles are segregated from each other when reproductive cells are made. Each F_1 plant will produce reproductive cells; half of the reproductive cells will have the tall allele and half will have the short allele. What would be the result when two F_1 plants are crossed (see Figure 7.5)?

Figure 7.5
Segregation of paired alleles occurs during gamete formation. The alleles are paired up again when gametes fuse during fertilization.

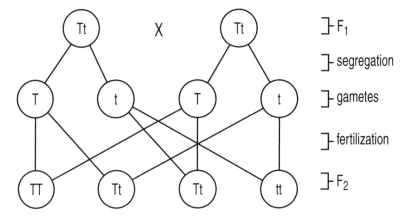

A *Punnett square*, like that shown in Figure 7.6, is a diagram which is used to help predict the results of crosses between organisms having variants of a trait to show more clearly the possible gene combinations resulting from a cross. The square (after fertilization) also shows each possible gene combination for the offspring in the boxes that make up the square. The letters on the top represent the gamete alleles for one trait for one parent and the letters on the left side represent the gamete alleles for one trait for the other parent. The case of the letter indicates whether it is a dominant or recessive allele.

To determine how the Punnett square can be used to predict outcomes of fertilization, follow steps 1–7 on the following page to predict the outcomes of Mendel's experiments by crossing a plant having two dominant alleles (*RR*) for round seeds with a plant having two recessive alleles (*rr*) for wrinkled seeds.

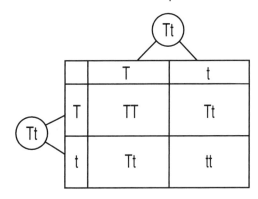

Figure 7.6
A Punnett square showing the cross between two heterozygous pea plants for stem height.

STEP 1. Make a key that represents the alleles in the cross.

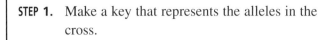

R = Round
r = Wrinkled

STEP 2. Write the genotypes of the parents.

RR x rr

STEP 3. Determine the possible gametes that each parent can form.

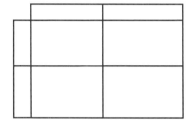

STEP 4. Create a Punnett square.

STEP 5. Enter the possible gametes from one parent at the top of the Punnett square, and those from the other parent on the side.

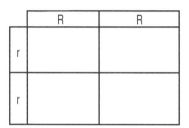

STEP 6. Complete the Punnett square by combining the gamete alleles in the appropriate boxes.

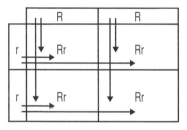

STEP 7. Compile the genotypes and phenotypes.

Phenotypes of the F_1 are all round.
Genotypes of the F_1 are all *Rr*.

    Are your results the same as Mendel's F_1 data? The predicted outcome was that all the peas would look round in the first generation. The square shows that when you cross two organisms that are homozygous (have two identical alleles for a particular trait)—one dominant parent and one recessive parent—all the offspring will have two alleles, a dominant and a recessive, for that trait (a *monohybrid* cross). That is, the offspring are all heterozygous. The square also confirms Mendel's results that for each cross of a homozygous dominant variant with a homozygous recessive variant, all the plants showed the dominant phenotype.

Does the data from Mendel's F_2 experiments fit this model? The predicted ratio—3 dominant phenotype to 1 recessive phenotype—showed up consistently in Mendel's results. What would a Punnett square look like if you crossed peas from the F_1 generation shown above? Would the predicted phenotypic ratio appear? How is the genotype related to the phenotype?

▶ ANALYSIS

Use the Punnett square as you respond in your notebook to the following.

1. In the fruit fly *Drosophila melanogaster*, wings (*W*) are dominant over a lack of wings (*w*), and red eyes (*R*) are dominant over sepia (brownish) eyes (*r*). A wingless fly with sepia eye color is crossed with a fly that has wings and red eyes (a *dihybrid* cross, which consists of variants in two traits). What are the possible genotypes of each parent? What are the possible genotypic and phenotypic ratios of the offspring? What fraction of the offspring from this cross might be wingless and have sepia eyes? What fraction will have the genotype *WwRr*?

2. You have a winged red-eyed fruit fly in your laboratory. Design a cross to determine whether the fly is heterozygous for both traits. Use a Punnett square to show all possible crosses.

3. Genetic counselors often use both Punnett squares and pedigrees to predict the probability of a trait variant occurring in a particular individual. Albinism is a recessive variant which results in a lack of pigmentation of skin and hair. Imagine that you are a genetic counselor with the following data:
   - Two couples—Harry and Theda T., and Jared and Emma A.—are normally pigmented.
   - Harry and Theda have two daughters who are normally pigmented.
   - Jared and Emma have four children, three sons and one daughter, who are normally pigmented.
   - Harry and Theda's second daughter, Sally, marries Jared, Jr., Jared and Emma's son.
   - Sally and Jared, Jr. have two children; their son is normally pigmented and their daughter lacks pigmentation in her skin and hair, that is, she is an albino.

Sally and Jared, Jr. wish to have another child and wish to know the probability of this child being an albino.
   a. Use the pedigree symbols found in Learning Experience 2 and create a pedigree of Sally and Jared, Jr.'s families.
   b. Make a Punnett square of a mating between Sally and Jared, Jr.

c. What is the probability that their third child will be an albino?

d. What is the probability that their third child will be a carrier?

4. The 23rd chromosome pair determines gender in humans. Using a Punnett square, determine the probabilities of a couple having a male baby (*XY*), and the probability of having a female baby (*XX*). A couple who has two daughters are planning to have a third child. What are the odds of this child being a boy? Explain your response.

A Growing Concern

Long before Mendel's work on pea plants, many farmers and gardeners were very interested in how traits were passed on from generation to generation in their livestock and crops. The practice of agriculture arose in several areas of the world, with the earliest evidence dating back approximately 10,000 years in the Near East. It is not clear exactly what factors led early people to change from a life of foraging to a life of cultivating their plants from seeds in gardens and farms. But whatever the causes of the change, the key discovery in this change of lifestyle was that plants could be cultivated from seeds. Farmers soon learned that by using only the seeds from crops that gave the highest yields, that were the most drought- or disease-resistant, that were easiest to harvest, or that tasted the best, they could improve the quantity and quality of each succeeding year's harvest. Farmers bred cereal grains such as wheat and rye for characteristics of hardiness, drought resistance, and yield. Gardeners were interested in creating plants with different colored flowers—combining variant forms of the same trait.

In more recent years, scientists have developed genetic engineering techniques designed to bypass the vagaries (chanciness) of inheritance of traits. These techniques enable them to insert genes from one organism into another organism in order to induce that organism to express new traits. Dozens of genetically engineered crops—such as Flavr Savr tomatoes, which have had flounder genes added to toughen up the fruit for shipping and to increase the shelf life, or soybeans, which have had genes from petunia, bacteria, and cauliflower added to improve herbicide tolerance—have already hit the market, and many more are on the way.

A great deal of controversy has arisen over the production and use of genetically engineered crops. Proponents claim that the ability to add specific characteristics to fruits, vegetables, and cereal grains, such as those added to the Flavr Savr tomato, will provide nutritious food to more people year round. The ability to make crops frost-resistant and herbicide-resistant may make food more affordable. Crops could be made resistant to damage by insects and fungi and to infection by bacteria and viruses, reducing the need for environmentally polluting pesticides and other chemicals.

Opponents, however, have a different perspective. Some members of consumer and environmental groups contend that it is impossible to predict the long-term consequences of the consumption of genetically engineered crops on human health. They worry that the insertion of genes into organisms in which they are not normally found may have serious consequences for the ecological balance of the environment, one which laboratory and field test trials cannot predict.

At the present time, the majority (about 93%) of genetic alterations are made to make food production easier and more profitable. The remaining 7% are engineered to improve taste or nutrition. What are the goals of modern techniques of creating new varieties of crops? Do the goals justify the means? The following article discusses the introduction of genetically engineered soybeans into the consumer market.

Gene Engineered Soybeans Get Frigid Reception

by Philip R. Reilly, M.D., J.D., in The Gene Letter, Volume 1, Issue 3, November 1996.

The first ship load of Monsanto's genetically engineered soybeans got a frigid welcome at the docks in Hamburg, Germany when it arrived in early November. Despite much scrutiny and regulatory approval from the FDA in the U.S. and the European Union, some food producers and supermarket chains have vowed they will not buy the beans or any product that contains them.

The fuss is over a single gene that Monsanto scientists transferred into the soybean plant so that it would be resistant to a major herbicide (also manufactured by Monsanto) called Roundup. Since Roundup kills all nonresistant leafy plants, the idea was to increase soybean crop yield by reducing weed encroachment.

Public opinion polls have showed that Europeans are in general more concerned about bioengineered food products than are Americans. The Green Party has led the way in the attack against genetically engineered agricultural products. More than 80% of Europeans want such products clearly labeled as having been bioengineered. It is difficult to know the extent to which the tremendous fear generated by "mad cow disease" in the United Kingdom is influencing the reaction to the soybeans, but [the fears] must be making consumers generally more suspicious.

What is clear is that agribusiness giants like Ciba-Geigy, DuPont, Dow and Hoechst have made huge commitments to developing genetically engineered crops that will help to answer humanity's ever growing need for food. While Monsanto's new soybeans represent the first large scale entry of a genetically engineered crop into the European market, there are more than 50 food products in the U.S. product pipeline. Plans to introduce genetically engineered corn, chicory, and rapeseed in Europe are well underway.

Since the soybeans have passed all the necessary regulatory hurdles, the battle is likely to be fought at the checkout counters. If products using the modified plant are cheaper, economics is likely to determine acceptance. That, surely, is what Monsanto is betting.

▶ ANALYSIS

Write responses to the following in your notebook.

1. Compare the similarities and differences between a soybean selectively bred for a certain characteristic and one genetically engineered for a certain characteristic.

2. If the food has passed the necessary regulatory hurdles, should the food be labeled "genetically engineered"? Why or why not?

3. Would you eat foods that have been altered with bioengineering techniques? Include your reasons for making the choice.

4. List four important values that influenced your decision.

▣XTENDING IDEAS

◐ Reports have surfaced in recent years that Gregor Mendel may have altered his data in order to have the numbers fit the expected monohybrid cross phenotype of 3:1 (dominant to recessive) and the dihybrid cross phenotype of 9:3:3:1. Research this controversy and state your opinion with your reasons.

◐ Five families, living in small villages close together in Colombia, have suffered as one member after another has fallen victim to an early onset of Alzheimer's disease. For generations, the families have regarded their affliction as a mystery, the subject of superstition. A team of Colombian and American researchers are creating pedigrees of the more than 3000 living family members. The cause of this form of Alzheimer's is a mutation of a single nucleotide of a gene (PS1) on chromosome 14. The dominant form, called E280A, has spread through the Antioquia families. Research how scientists are solving this mystery, identifying the gene, and how the gene codes to produce the Alzheimer condition. (For resources, see "Five Families' Anguish May Aid in Alzheimer's" by Richard Saltus, the *Boston Globe*, March 17, 1997, pages C1 and C4; and "Clinical Features of Early-Onset Alzheimer Disease..." by Francisco Lopera, et al. *Journal of the American Medical Association* (*JAMA*), March 12, 1997, pp. 793–799.)

▮❶N THE JOB

NURSERY WORKER   It was the middle of February in New England, but plants were flourishing in the warm, moist air at the Mountainview Nursery. Alan watered the orchids while his partner planted seeds for flowering perennials that would be sold as seedlings during the spring planting season.

Alan grew up in gardens. As a child, he would help his father with his landscaping business. He enjoyed working with his hands and learning about all of the different kinds of plants and the best conditions for growing them. As he grew older, he found he was very interested in the nurseries that supplied his father with different species of plants, which he then arranged in customers' yards and gardens. During Alan's high school years, he spent his vacations working in a neighborhood nursery. He would water the plants, check them for insect damage or disease, prune them to create fuller foliage and more numerous blooms, and collect seeds for the next year's crop. He always felt proud when one of his seedlings or adult plants was sold. As a plus to his job, during the winters—when everyone else was complaining about the cold and the snow—he was always able to escape to his warm haven and surround himself with petunias, dahlias, azaleas, and roses, with greens like ivy, coleus, and ferns, and with many varieties of cacti.

As an adult, Alan owns his own nursery. He periodically takes courses to learn new techniques of nursery work (what different seeds are available, what technology exists for temperature control, watering, pesticides, etc.) and has also taken numerous college courses in business management, horticulture, and biology. But his experience while growing up was the most important part of his training.

Besides maintaining and propagating small plants, shrubs, and trees, Alan keeps up a stock of gardening supplies for the nursery and for his customers. Other stock includes bird seed and feeders, because birds are considered an important ingredient in a successful garden. Alan follows gardening trends, watching out for what plants are particularly popular during any year and bringing in unique plants from around the world.

Alan recently made an addition to the nursery: a section for aquatic plants. Many of his clients have small ponds in their gardens and want to increase the amount of plants in and around them. His new addition includes water lilies, blue flag irises, and several species of tall grasses.

In order to share all of the knowledge he has acquired over his years in the gardening and nursery business, Alan teaches regular seminars in his store. And he has made a rule of hiring a few high school students each year, hoping to propagate his love and appreciation of the natural world through new generations.

Mapping Genetic Trails

PROLOGUE   **M**endel's observations of patterns in the inheritance of traits in peas led him to inferences that were later confirmed by research on chromosome movement during meiosis. Mendel inferred that alleles separate and segregate during gamete formation. He also thought that a gene for one trait did not influence the inheritance pattern for a different trait—that is, that genes demonstrated independent assortment. However, if you think about the number of chromosomes in relation to the number of genes an organism has, there seems to be a problem. A human, for example, has approximately 80,000 different genes, all of which are located on 23 pairs of chromosomes. Therefore, there are thousands of genes on each chromosome (represented by the marker bands on your chromosome models). The problem then is, if two genes are located on the same chromosome, can they sort independently? If not, does this contradict Mendel's principle of independent assortment?

In this learning experience, you will investigate how physically linked genes sort. You will also explore how scientists are using technologies developed from their understandings about chromosome segregation and assortment in one of the major research efforts of the twentieth century—the Human Genome Project.

I'll Trade You

INTRODUCTION   If segregation and independent assortment occur only in the ways Mendel proposed, then the possible gene combinations in the offspring should be predictable. However, geneticists soon discovered the concept of *gene linkage*, that all of the genes located on the same chromosome move together when that chromosome segregates during meiosis.

The fact of gene linkage seems to limit variation. What else might occur during meiosis that would increase variation? In this activity, you will examine the chromosomes found in gametes following meiosis and apply Mendel's principles to analyze the combination of alleles in the gametes.

▶ MATERIALS NEEDED

For each pair of students:
- colored pencils or felt-tip markers–yellow, gray, brown, green, blue, red, and black

▶ PROCEDURE

1. A mythical mosquito has the following genotype:

   Hh Bb Rr Ll

   By chance, all of the dominant alleles came from the mother (maternal chromosome) and all of the recessive alleles came from the father (paternal chromosome). Use the colored pencils to create a drawing of the three chromosome pairs of your mosquito. Place the alleles in the appropriate locations. (The chromosomes and the location of the alleles on these chromosomes should be based on the model from Learning Experience 6. Draw one of each set of homologous chromosomes, one yellow representing the paternal chromosome and one white representing the maternal chromosome.)

2. Determine the possible gametes for your mosquito and draw the chromosome contents of the gametes.

3. Use Table 8.1 to write responses to the following in your notebook.
   - If a gamete of this mosquito contains information for a black abdomen (B), which allele of wing shape would it have—R or r?
   - If a gamete of this mosquito contains information for an orange abdomen (b), which allele of wing length would it have—L or l?

Table 8.1

TRAIT	PHENOTYPE	SYMBOL FOR ALLELE	BAND COLOR	CHROMOSOME (cm)
antenna texture	hairy smooth	H h	red	15
wing shape	round square	R r	green	30
wing length	long short	L l	black	22
abdomen color	black orange	B b	blue	30
unknown	not available	not available	brown	on all chromosomes

4. While examining this mythical mosquito, a geneticist observed the following genotypes:

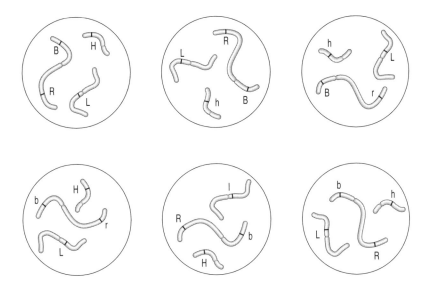

Figure 8.2
Some observed genotypes of gametes in the mythical mosquito.

Describe the differences between the gametes you predicted in step 2 and the gametes in Figure 8.2.

5. In a short paragraph or a drawing, propose an explanation that could account for the unexpected genotypes.

HAIRY, THE BLUE TOMATO

INTRODUCTION The Jolly Blue Tomato Company specializes in producing novelty fruit. They have found that consumers have become daring in their tastes and crave new and exotic fruits and vegetables as a way of meeting the daily intake of five fruits and vegetables recommended by the health specialists.

Your team at Jolly Blue has been assigned the task of breeding a tomato which is blue and hairy and grown on a dwarf plant, so that many plants can be grown in a small space.

▶ TASK

1. Your parent plants have the genotypes *Ss, rr, TT* and *ss, Rr, tt*. You also know that all three genes are located on the same chromosome. Use Table 8.3 to describe the phenotypes of your parent plants.

2. Your team has carried out many, many crosses and examined hundreds and hundreds of plants and their fruit. You have been able to produce plants and tomatoes with the following characteristics:

Table 8.3

TRAIT	PHENOTYPE	SYMBOL
color	red	R
	blue	r
texture	smooth	S
	hairy	s
height	tall	T
	dwarf	t

- blue, smooth, and tall
- blue, smooth, and dwarf
- red, hairy, and tall
- red, hairy, and dwarf

However, you have not been able to produce a plant with blue, hairy, and dwarf characteristics. Your team needs to write a report for the president of Jolly Blue Tomato Company (who feels anything is possible), describing what needs to happen in order to achieve the requested results. Your team is familiar with the modes to achieve variation, including that of *crossing over* in which there is an exchange of alleles between homologous chromosomes. Based on the teams understanding of variation, include in the report a hypothesis which might explain why producing a hairy, blue tomato may **not** be possible by conventional breeding techniques.

HOME, HOME ON THE CHROMOSOME

The human *genome* consists of all the genetic material present in a human's chromosomes. Even though it is estimated that only about 5% of human DNA contains information for coding proteins, even this small percentage consists of approximately 80,000 genes.

Within this vast amount of DNA, how is it possible to determine the location of a single gene which may be no bigger than a few thousand bases? Finding the location of specific genes on specific chromosomes has been the goal of many researchers of genetics. But this task is akin to locating individuals in the United States when you don't know their address, their city, or even their state of residence. Both tasks take a great deal of creativity, ingenuity, detective skills, plain hard work, and lots of money.

HOW DO YOU FOLD A GENETIC MAP?

The first gene to be located on a chromosome (*mapped*) was the gene responsible for red-green color blindness. It had been observed that the inheritance of this disorder was passed from mothers who were able to perceive colors normally only to their sons. As you will recall from Learning Experience 5, females have *XX* sex chromosomes and males have *XY* sex chromosomes. Although genes occur in pairs on the 22 pairs of *autosomes* (chromosomes other than sex chromosomes). There is little or no homology between the *X* and the *Y* chromosome. Therefore, a gene on the *X* chromosome (even a single recessive one) will be expressed. Females, however, with a normal allele on one *X* chromosome and a recessive allele on the other *X* chromosome would not show the recessive trait. This logic enabled researchers to map the gene for red-green color blindness and several other genes (including the gene for the blood clotting disease hemophilia) also on the *X* chromosome. Genes located on the sex chromosomes are known as *sex-linked genes*, and the appearance of the associated phenotypes depends on the individual's gender.

In the 1960s, identification of genes on the other 22 chromosomes began. Early mapping techniques involved joining or fusing in a test tube mouse and human cells into a new kind of hybrid mouse/human cell. This fusion had an interesting result: The newly created hybrid cell literally tossed out its human chromosomes until only a few remained. By determining what human proteins still were being made in the hybrid after this chromosome-clearing process, scientists could assign the genes for these proteins to the human chromosomes that were still present in the hybrid cell. This approach, coupled with new staining techniques, enabled scientists to assign about 1,000 genes to specific chromosomes.

More precise approaches to gene mapping came with two central observations. The first was that genes on chromosomes tend to be inherited together unless crossing over occurs. The distance between two genes can be determined by the frequency of crossing over; that is, how often two genes on the same chromosome are separated from one another. The closer two genes are, the less frequently they will be separated by crossing over. Traits that often occur together, such as freckles and red hair, most likely are the result of gene linkage and are located close to each other on a chromosome. Genes that are linked show a very low frequency of crossing over. Figure 8.4 gives an example of the relationship between the frequency of crossing over between alleles and the location of the genes on a chromosome.

The second observation, made in the early 1980s, was that individuals not only look different from one another, but also have distinctive chromosomes. Some of these distinctive features appear as variations in chromosome staining patterns (patterns of the DNA-binding proteins); others are variations in the DNA sequence which may occur as frequently as every 500 bases, primarily in stretches of DNA that have no coding function.

These regions of variation, called *genetic markers*, can be used to follow inheritance of specific chromosomes or regions of a chromosome through many generations of a family. By following the inheritance of these markers and the inheritance of specific genetic diseases,

a)

Maternal Chromosome

	A	B	C	D	E
A	0	.25	.10	.04	.41
B	.25	0	.31	.22	.60
C	.10	.31	0	.03	.32
D	.04	.22	.03	0	.35
E	.41	.60	.32	.35	0

Paternal Chromosome

b)

B   A D C   E

Figure 8.4
Example of linked genes and the frequency of crossing over. (a) The numbers in the chart indicate the frequency of crossing over. The genes that are farthest apart, for example *B* and *E*, demonstrate greatest frequency of crossing over (.60). (b) The location of these genes on the chromosome, as determined by frequency of crossing over.

scientists have been able to find the chromosomal locations of genes responsible for disorders such as cystic fibrosis, sickle cell anemia, Tay-Sachs disease, and fragile X syndrome. By 1994, a map of the human genome locating 5,000 markers and more than 400 genes was created.

Once this kind of genetic linkage map is used to locate the gene to a relatively small area of the chromosome, scientists then construct a physical map which shows the actual position of a gene along the chromosome. The ultimate goal for creating physical maps of genes is to determine the DNA sequence.

THE HUMAN GENOME PROJECT

One of the major scientific initiatives of the twentieth century is to map the approximately 80,000 genes to their locations on the 24 human chromosomes (22 autosomes and the *X* and *Y* chromosomes). This effort, known as the *Human Genome Project*, hopes to create complete genetic and physical maps of the human genome early in the twenty-first century. Its goal is to identify all of the instructions encoded in our DNA and to read the language of our chemical inheritance. Maps developed through the Human Genome Project will enable researchers to locate specific genes on our chromosomes and to understand how these genes function or, in some cases, fail to function.

Scientists hope that by having a complete genetic and physical map of the human genome, many medical problems can be solved and basic scientific questions answered. The potential medical benefits are enormous; by knowing the sequence of specific genes and the proteins that they encode, scientists can develop new tests for detecting disease, design new drugs for treatments, and perhaps reach the ultimate goal of replacing faulty genes with normal genes (*gene therapy*). By becoming aware of a predisposition to certain diseases through genetic screening, individuals might be able to delay or prevent the onset of symptoms through changes in their life style and habits. In addition, scientists could determine how specific environmental factors such as chemicals, drugs, and pollutants affect and alter genes.

Enormous benefits of unraveling some of the mysteries of human history and existence may also result. Sequencing the entire human genome and comparing it to the genomes of other organisms can clarify evolutionary relationships among different species. Tracking the appearance of mutations will help establish a more precise evolutionary timeline.

With a map of the human genome in hand, scientists may be able to determine the mechanisms that control gene expression and the processes that regulate the careful timing of a cell's growth and division (and how this may go out of control in cancer). Identifying certain genes and understanding how they are expressed can also lead to developing new strains of fruits and vegetables and new breeds of animals. Finally, it is hoped that by deciphering the human genome, we can understand one of life's greatest mysteries: How does a single fertilized egg grow and develop into a multicellular organism?

Although the Human Genome Project seems to promise a great deal in regard to curing disease and understanding many questions in biology, many fear the consequences of such a project. From its beginning, the costs of the project have been enormous in terms of money (more than three billion dollars from U. S. taxpayers), time, and numbers of individuals involved. Many feel that this is not a wise use of the taxpayers' money and that it is diverting resources from other important research.

A major ethical concern is that once the human genome is sequenced and genes identified, information about an individual's genetic makeup could be used against him or her. Following analysis of a drop of blood or a snip of hair, an individual's deepest genetic secrets might be revealed to anyone caring to know them. Bioethical questions have already arisen as genes for genetic disorders are identified and diagnostic tests made available. Is it appropriate to diagnose a disease for which no cure is known? Examples of this include sickle cell anemia and Huntington's disease. Will employers or insurance companies use these test results against the individual?

It is conceivable that one day every individual might carry a DNA identification card rather than a driver's license. This card would identify an individual more precisely and thoroughly than any description or fingerprint. Who should have legal access to this card? Is this something you would feel comfortable having?

▶ ANALYSIS

Write responses to the following in your notebook.

1. Using your understanding of frequency of crossing over as a way of determining gene location on a chromosome, reconsider your explanation of the failure to create a hairy blue tomato in the activity "Hairy, the Blue Tomato." Decide where the alleles for the traits S, T, and R might be located on the tomato chromosome and indicate this on a diagram.

2. You are an aide to a congressperson and have been assigned the task of preparing him/her for a vote in Congress on whether funding for the Human Genome Project should be continued. You need to write an informative position paper that can be used during a debate. Your paper must include:
    *   the goals of the Human Genome Project
    *   an explanation of the scientific concepts involved in the Project
    *   the potential benefits and abuses of the information derived from the Project

- the economic, legal, and ethical issues involved
- your recommendation as to how he/she should vote and why

Your letter must be brief, concise, and very clear, as your congressperson has neither a strong understanding of biology nor much time to prepare for the vote.

EXTENDING IDEAS

▶ The distinctive characteristics of each individual's chromosomes have provided defenders of the law with a new weapon in the fight against crime. Known as *DNA fingerprinting*, this technique relies on the observation that the DNA of every individual creates a unique pattern when treated with enzymes to cut the DNA into discrete fragments that can be visualized by various methods. This technique has been used to match the DNA of suspects with DNA isolated from blood or hair found at the scene of a crime. However, some critics question its validity and reliability. Research this technology and describe the principles behind it. Identify any high-profile cases in which it has been used and the arguments for and against its use. Using your understanding of the science of DNA fingerprinting, explain which side you support and why.

▶ Treatment of genetic disorders includes recent technology in gene therapy. Two approaches are possible. In one instance the appropriate gene is delivered to the specialized cells in which the altered protein is expressed; for example, in sickle cell, a functional ß-globin gene would be delivered to cells which eventually form red blood cells. This is called somatic gene therapy. Another possibility is to deliver the gene to the germline cells or gametes. In this approach the altered gene would be "fixed" for all the progeny of that individual. Describe the differences between somatic and germline gene therapy in terms of long range effects for both the individual and society. Serious concern has been expressed about the potential uses and misuses of germline gene therapy. Research the ethical issues involved in this technique and present both sides of the controversy.

▶ Gene therapy has been used in trials to treat the genetic disease cystic fibrosis. Research the molecular basis of cystic fibrosis and current attempts to treat it by gene therapy.

DATABASE ADMINISTRATOR   The door of the computer laboratory opened, and Dr. Nadeau approached Mitra to discuss a presentation he was preparing, for which he needed her input. He was doing the presentation for an upcoming conference. The topic was on a recent cloning experiment his laboratory had attempted. This particular experiment was not successful, but Dr. Nadeau wanted to share the results of his research with the group of colleagues attending this conference. He needed Mitra to organize the data collected from this experiment and to illustrate the new knowledge obtained from the research.

Mitra's work is essential to the scientists with whom she works. She is often asked to run statistical analyses on the data and graph the results. Mitra is also responsible for keeping the database up to date and in good working order. In order to do this, she relies on her attention to detail. Because this is a highly competitive and fast-paced field, Mitra needs to keep an eye on the security of the entire database system. There are people who, before the scientific data is published, would be very interested in having access. Mitra had to make sure that access is available only to those scientists and technicians directly involved in the research.

Mitra learned computer basics in middle and high school. She always enjoyed working with numbers and data, and found it easy to use computers for performing the tasks she had in mind. Her first job after high school graduation was in a physician's office where she was exposed to medical terminology. Becoming interested in biology and medicine, she enrolled in some undergraduate biology courses at a community college. Her computer background and willingness to learn more about biology, along with her experience in her first job, makes her present position perfect for her. In addition, Mitra must keep up with the latest technological advances in computers and observe how technology is ultimately affecting the field of genetics.

When Dr. Nadeau returned from the conference, he said the tables and graphs she had created for his presentation were very well received. Several scientists commented that the results were extremely clear. These results will be published in a medical journal and then, perhaps, referred to by others for lobbying purposes or in courts of law. The news gave Mitra a strong sense of accomplishment and pride.

WHAT MENDEL NEVER KNEW

PROLOGUE   Look around you. The individuals in your class are a living illustration of how phenotypes are not always as straightforward as in Mendel's peas. Mendel made several fundamental observations about inheritance patterns in peas. He concluded that traits, such as the height of a pea plant, were determined by "discrete factors" (genes) that occur in pairs; one member of each pair was inherited from each parent. Mendel's crosses yielded offspring that were easily distinguished from each other. A plant was either tall or short and the variant of tallness would dominate over the variant for shortness. These straightforward patterns are known as "Mendelian genetics" or Mendelian inheritance.

As later scientists investigated the patterns of inheritance in many different organisms, they discovered that Mendelian genetics was not sufficient to explain some phenotypic variations. In humans, dominant/recessive gene interactions are not frequently seen. For example, if one parent is tall and the other short, the phenotypes of their offspring do not follow the simple pattern of all tall or all short. The pattern is much more complex. In this learning experience, you will explore traits which result from a variety of gene product interactions.

VARIATION: IT'S NOT THAT SIMPLE

INTRODUCTION   In many cases, several genes may contribute to the phenotype; the products of these genes (the protein encoded by the gene) are neither dominant nor recessive, and the interactions produce a range of variations in the trait. You are going to begin an analysis of several traits in order to explore various modes of gene interactions. Through the descriptions and Table 9.1, you should be able to determine how these interactions help explain the variation of traits observed in many organisms.

▶ TASK

1. Read each of the descriptions of traits that follow Table 9.1.
2. Use Table 9.1 to identify which mode of inheritance each trait exhibits.
3. Write responses in your notebook to the statement found at the end of each trait description.

Table 9.1
Modes of gene product interaction.

▶ MODE OF INHERITANCE	DESCRIPTION OF INTERACTION RESULTING IN PHENOTYPE
Complete Dominance	A gene has two alleles, which may encode variant forms of that protein resulting in different phenotypes. When the two different alleles are present, only one phenotype will appear (dominate).
Incomplete Dominance	A gene has two or more alleles, and the phenotype is the result of the interaction of the variant products of both alleles. The phenotype may appear as a blending of the two products although the alleles continue to separate independently.
Codominance	A gene has two alleles, and the phenotype is the result of the action or interaction of both variant products of the alleles. This is similar to incomplete dominance, except that the products are discrete rather than blended.
Multiple Alleles	A gene has more than two possible alleles; they all encode variants of the same protein. The phenotype is dependent on which two alleles are present in the organism and the patterns in which the products of these alleles interact. The patterns may be dominant, incompletely dominant or codominant.
Pleiotropic	A single gene may have multiple effects on the phenotype of an organism. Both alleles of the gene encode a protein with an altered function; the failure to produce a functional protein alters many characteristics of the organism.
Polygenic	The phenotype is the cumulative result of the interactions of the products of several genes and their alleles.

TRAIT 1

Sickle cell anemia is the result of a change in the ß-globin chain of hemoglobin. An individual who is homozygous for sickle cell trait will show classic symptoms of sickle cell anemia; intense joint pain, shortness of breath, anemia, and the characteristic sickle shape of the red blood cells. In an individual who is heterozygous for sickle cell, the alleles for both types of hemoglobin (A and S) are present in every red blood cell, but the sickling phenotype only appears under conditions of oxygen deprivation, such as during exercise at high altitudes.

Identify the mode of inheritance and explain the phenotype of individuals who are homozygous, and those who are heterozygous for this gene.

TRAIT 2

The so-called "blue" (actually gray) Andalusian variety of chicken is produced by crossing a black parent and a white parent. Color production in these chickens depends on a single gene. In the F_2 generation, black and also white chickens may reappear.

Identify the mode of inheritance and describe how this gene product interaction produces the observed phenotype.

TRAIT 3

Human height is determined by a number of factors, including diet. But even if all individuals were fed the same diet, height among individuals would show continuous variation, that is, a gradation of small differences within a certain range. Height is determined by a number of gene products including levels in production of hormones such as growth hormone, and growth capacity of structural components such as cartilage, connective tissue, skeletal muscles, and bone.

Identify the mode of inheritance for the trait of height and explain how this type of gene product interaction contributes to the variation of height.

TRAIT 4

The color of human skin is determined by genetic factors, but the shade may vary from environmental factors. Skin color is almost entirely the result of the amount of melanin pigment and its distribution in the outer layer of skin. At least two genes have been identified as being involved in melanin production; two genes produce large amounts of melanin and the alleles of these genes produce small amounts of melanin. Several other genes appear to be involved in how this melanin is distributed in the skin. Melanocytes, special cells which produce melanin, are much larger in darker-skinned individuals, and have many more specialized structures which deliver the melanin to the outermost layer of skin cells.

Identify the mode of inheritance and describe how this kind of gene product interaction might produce gradations in skin coloring.

TRAIT 5

In humans, the four major blood groups (A, B, AB, and O) are determined by a gene which encodes an enzyme used in the synthesis of a polysaccharide (large sugar molecule) found on the surface of red blood cells. This gene has three alleles, each of which encodes a variant of this enzyme; the variants result in the synthesis of the different forms of the polysaccharide which characterize the blood as being type A, B, AB, or O. Table 9.2 lists the phenotype (blood type) of an individual and the genotypes possible.

For each genotype, identify the mode of inheritance and describe how type of allele product interaction produces the resulting phenotype.

Table 9.2

Phenotype (Blood type)	Genotype (Alleles present)	Polysaccharides on Surface of Red Blood Cell
O	OO	
A	AA, AO	
B	BB, BO	
AB	AB	

TRAIT 6

A single gene in rats controls the production of a protein involved in forming cartilage, the tough elastic tissue that occurs in vertebrate animals and provides some of the organism's structural support. A rat carrying two alleles with an altered protein displays a whole complex of birth defects, including thickened ribs, a narrowing of the passage through which air moves to and from the lungs, a loss of elasticity in the lungs, blocked nostrils, a blunt snout, and a thickening of the heart muscle. These effects generally result in death.

Identify the mode of inheritance and explain the effects of the gene products.

PLAYING YOUR HAND

Variation can be explained by the processes you have been exploring in the last several learning experiences. Mendel's principles of segregation, independent assortment, and dominance, as well as the various modes of interactions of gene products, can account for the tremendous variation observed among individuals. Once you have been dealt your DNA, is the outcome of the game predictable? What role does environment play in the phenotypic expression of a genotype?

THE ONE AND ONLY YOU

The debate concerning nature vs. nurture—about the roles that genetic inheritance and the environment play in determining who a person is and will become—is an old one. At the beginning of the twentieth century, scientists were concluding that heredity is not destiny. Using daphnias, small freshwater organisms, it was demonstrated that organisms with the same genetic makeup, raised under different conditions of temperature and acidity, displayed varying physical and behavioral characteristics. The phenotypic expression of their genotypes was influenced by their environment.

Identical twins separated at birth have provided fascinating insights into the roles played by heritable and environmental factors. These individuals, produced from a single fertilized egg, share the same genetic makeup. Since 1979, traits of separated twins have been tracked in a study conducted at the University of Minnesota. In these studies, physical and behavioral traits, including handedness, fingerprint pattern, height, weight, intelligence, allergies, and dental patterns were measured. The twins' interests, fears, habits, and beliefs were compared. The researchers found that identical twins separated at birth and reunited later were remarkably similar, right down to certain twitches and idiosyncrasies. However, enough variations in traits were also observed to indicate that the environment also makes a contribution.

For certain traits, the effect of environment is readily apparent. In the past 100 years, the average height of humans has increased, mainly due to better nutrition. Although genetic inheritance plays a major role in determining height, diet can influence the final outcome. In Himalayan rabbits and Siamese cats, the fur on their extremities—ears, nose, and paws—is much darker than the fur on the rest of their body. This is due to a heat-sensitive allele of a gene that encodes an enzyme which influences melanin production. In warm body regions, the enzyme is less active, and so fur grows in lighter than it does in the cooler tips of the body.

For other traits, however, the link is less obvious. Imagine that at birth you had been moved to an environment very different from your original one. How might you be different now? Would your genetic inheritance continue to make you "you" or would your environment have influenced you to become a very different person?

GENETIC JEOPARDY

The relationship between having a gene and how it is expressed is not as clear-cut as it seems and this is seen most clearly in inherited diseases. The locations of genes for many heritable diseases are being mapped on chromosomes through the efforts of the Human Genome Project; some of these are the genes for cystic fibrosis, Huntington's disease, sickle cell anemia, and breast cancer. With this information, genetic counselors can, in theory, identify those individuals who have the gene and, therefore, may have a genetic predisposition to the disease. But is having a faulty gene the same as having the disease? In cases such as sickle cell and cystic fibrosis, having one allele which encodes an abnormal protein does not result in disease; these are cases where the abnormal protein product is recessive. In other cases, such as Huntington's disease, the abnormal protein is dominant and, having a single copy, even in the presence of a normal copy of the gene, means that the individual will be affected.

Because a trait can be the result of the interactions of several alleles or gene products, the outcome of having an abnormal gene may not be so easy to predict. Does having the breast cancer gene mean you are destined to develop breast cancer? Can maintaining a certain life style reduce the risk of developing breast cancer? While the answers are not clear at this time for some diseases, for other diseases much is known.

The cardiovascular disease hypertension appears to have both genetic and environmental causes. During circulation of blood, the heart must pump blood against pressure in the arteries. Normally the arteries are flexible and present little resistance to blood flow. However, as the pressure in the arteries rises (for example, when fat deposits clog the arteries), the work that the heart must do increases. In an individual with a high blood pressure (180/140), the heart must work twice as hard as the heart of an individual whose blood pressure is 100/70. The consequences of this extra work for the heart can be serious; the heart muscle can be damaged, putting the individual at risk for a heart attack. Twenty to fifty genes may be involved in regulating blood pressure. One of the genes known to be involved encodes a protein, angiotensinogen, that controls blood vessel tone and flexibility, and some individuals with hypertension show variants in this protein.

The interplay between genetic and environmental factors is even more complex in mental and behavioral traits such as alcoholism, intelligence, sexual orientation, anxiety, and mental disorders. Headlines routinely proclaim that the genes for certain personality traits have been discovered; even if these genes exist, unraveling the complexities of their interactions with other gene products and the environment will be a challenge for geneticists.

▶ ANALYSIS

Write responses to the following in your notebook.

1. The following list describes factors that appear to increase the risk of hypertension:
    - diet high in fat
    - diet high in salt
    - stress
    - family history
    - insufficient exercise
    - obesity
    - maleness
    - African ancestry
    - alcohol intake
    - age

    Your latest medical examination indicates that your blood pressure is 190/140. Your doctor wants to know about the occurrence of hypertension in your family and would like you to change some of your habits.
    a. Explain why your doctor thinks it important to have information about your family.
    b. Describe what steps you can take to control your blood pressure.
    c. Do you think these steps will help even if you have a family history of hypertension? Why or why not?

2. What are some human traits or diseases that you think are the result of both genetic inheritance and environment? Explain your reasoning. Pick one human trait and design an experiment which would provide evidence to test your hypothesis.

Ever since the decision was made to attempt to identify the approximately 80,000 genes in the human genome, the rate of discovery for individual genes has climbed steadily. Scientists have discovered an obesity gene that helps regulate weight, genes that cause or contribute to cancer, and a gene that triggers the onset of Alzheimer's disease. Other research has suggested that traits that define important parts of our identity—whether we are happy or sad, predisposed to alcoholism or schizophrenia—may also be written in our DNA. Simultaneously scientists have uncovered some of the mechanisms by which the environment can influence genetic destiny by "turning on" or "turning off" various genes. This seems to strengthen the notion that nature and nurture are inseparable and that genes by themselves do not determine one's medical or behavioral fate. Research both sides of the debate and determine whether you feel we are who we are because of the genetic traits passed down through the generations or whether environmental influences can and have altered our genetic destiny.

ON THE JOB

ARCHAEOLOGIST   The day has come. It is Ibrahim's first day working for the Cairo Museum. His supervisor leads him to a heavy, locked door at the end of a long corridor. Ibrahim walks in and there she was. He's waited all these years. Mummy #46 lay on the table. Her wrappings are barely clinging to dusty hands, her bony fingers protruding. She is awaiting his analysis. Ibrahim hopes to learn as much as he can about her, while showing her the utmost respect, for she lived thousands of years ago.

Ibrahim grew up in the shadow of the pyramids with a dream of studying the pharaohs (kings) and queens who once ruled over his homeland. While in school, Ibrahim learned that over 500 mummies and mummy fragments were rediscovered in storage at a Cairo university and moved to the Cairo Museum to be studied. Ibrahim was determined to be a part of the study to identify these mummies, some of whom could have been a part of an Egyptian royal family.

Ibrahim made sure that along with his general archaeology studies, he also concentrated in genetics. This gave him the background he needed to work directly with the unidentified mummies. In his studies, Ibrahim found that in the past, x-rays of skull measurements were used to establish relationships between mummies. But as the technology of DNA collections and testing becomes more and more advanced, it gives the best hope of successful identification.

In school, Ibrahim took classes in archaeology and anthropology. In addition, Ibrahim was able to travel around the world to take part in a number of different archaeological digs. Some involved excavating sites of ancient civilizations. At each site, Ibrahim and other students and scientists studied the architecture, pottery, furniture, artwork, and human remains in order to make a hypothesis about what life was like when that civilization thrived. The field experience was very important.

During his travels, Ibrahim learned that archaeologists are also very interested in plant and animal life that surrounded the different sites of civilizations. Knowing what plant and animal life flourished in the surrounding regions helped his team learn about what the citizens of these communities used for food, clothing, and even shelter. For example, if they learned of a great drought causing devastation to crops and farming, Ibrahim's team could possibly conclude that starvation brought the metropolis to a devastating end.

After receiving a bachelor's degree in archaeology, Ibrahim chose to continue his education with a master's degree. His internship with the Cairo Museum gave him much hands-on experience, but until this point, he was not able to work with the unidentified mummies.

He studied her carefully. Many hours of analysis lay before him—tissue scrapings, x-rays, close examination of her teeth, the sparse hairs on her head, and even the position in which she lay. All of this would hopefully lead him to an identity, and finding one clear answer was sure to result in more as he continued his research within the maze of expressionless, nameless mummies.

GENETICS AND EVOLUTION: THE NEEDLE AND THE THREAD

PROLOGUE   **A**s you have seen, there is enormous variation in traits of humans and other organisms, owing to continual mixing and reshuffling of alleles during reproduction. But have there been genetic changes in a population of organisms since it first appeared on Earth? What might have caused these changes? Questions such as these link genetics to evolutionary theory.

What is the process of evolution, and how is it related to the principles of genetics? Evolution is often most simply defined as change in organisms over time. You are familiar with the fact that living things and structures change over time: people age, bicycles rust, and wood rots. But there are other types of changes: computers become faster and more powerful, astronauts stay in space longer, even years at a time, and diseases are prevented or cured. Change is an inevitable part of life and evolution is the concept that unifies the study of life.

In this learning experience, you will first explore the premises upon which Charles Darwin based his theory of evolution by natural selection and then learn how the discovery of DNA and the modern principles of heredity help to explain the scientific principles underlying Darwin's premises. You will be introduced to the field of population genetics, which synthesizes Mendel's principles, Darwin's theory, and current genetic knowledge.

ONWARD AND UPWARD?

INTRODUCTION Fossils—defined as any recognizable evidence of ancient organisms—open windows into the past, but also present scientists with several puzzles. When fossils of previously unidentified organisms are found, scientists attempt to interpret what the fossils reveal about the history of life on Earth. How complete is this new evidence? How does it fit in with what is already known? What does it mean in and of itself? The fossil record of life on Earth is incomplete, creating challenges to scientists who would like to chart the emergence and extinction of all living things in the 4.5 billion years since the formation of the earth. To their delight, "new" organisms are always coming to light, whether they are recognizable ancient remains or organisms living now but never before seen.

In this activity, you will create an *evolutionary tree*, that is, a logical pattern illustrating how a mythical organism known as "The Swingette" might have evolved over millions of years. Keep in mind that evolution is driven by randomness and chance, and that there may be "dead ends" in the evolution of this creature.

▶ MATERIALS NEEDED

For each pair of students:
- 1 sheet of fossil drawings— "The Swingette"
- 1 large sheet of paper (11x17 inches or larger)
- scissors
- 1 glue stick or roll of cellophane tape

CAUTION: Scissors may be sharp; handle with care.

NOTE: There may be extinctions or gaps in the fossil record.

▶ PROCEDURE

1. Cut out the 12 fossil drawings. You do not need to follow the lines exactly.

2. Place the fossil drawings on the large sheet of paper. Arrange them according to how you think this organism evolved over time.

3. Attach the fossil drawings to the paper either with a glue stick or tape and draw arrows between the drawings to create an evolutionary tree.

4. Add notations of your reasoning for the placement of the fossil drawings.

▶ ANALYSIS

Write responses to the following in your notebook.

1. What general pattern of evolution did you choose? Why?

2. Do you think evolution always occurs in this way? Give an example of how it might not.

3. What might be some reasons for an organism to become extinct?

4. What is the value of fossils to scientists if they do not give us a complete picture?

Charles Darwin's Theory— Then and Now

After graduating from the university, Charles Darwin (1809-1882) set sail on the H.M.S. *Beagle* in 1831 as both an unpaid dinner companion to the captain, who otherwise would dine alone, and as a naturalist. The mission of the *Beagle* and its crew was to chart the coastline of South America and continue around the world; it would be nearly five years before they returned to England (see Figure 10.1). At each stop along the way, Darwin went ashore, collecting and classifying organisms that few Europeans had ever seen. He experienced his first earthquake and climbed the Andes mountains where he found fossils of shellfish. He was intrigued by the finches on the Galapagos Islands, a series of islands of volcanic origin 600 miles off the shore of Ecuador in the

Figure 10.1
Voyage of the H.M.S. *Beagle.*

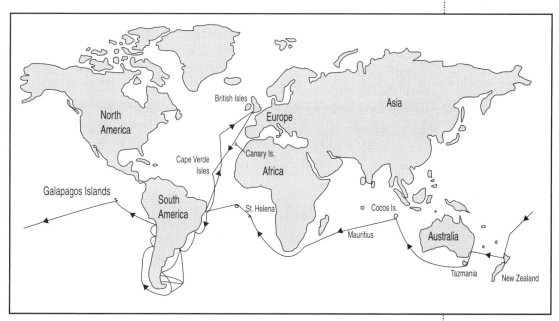

Pacific Ocean. He reasoned that all organisms on the Galapagos (including the finches) had to be "immigrants," brought to the islands by wind or currents or on floating debris. On each island there was a different species of finch; the beaks of each species varied in shape and size. They, in turn, were different from the original finch species that immigrated from Ecuador. What to make of these observations?

DARWIN'S EXPLANATION

Upon Darwin's return to England in 1836, he analyzed his vast quantity of data and specimens, read widely about populations and geology, and consulted with other scientists. For the next 20 years, he continued to define and refine his conclusions. In 1859, he published *On the Origin of Species by Means of Natural Selection*, setting forth his ideas on the evolution of organisms and changing the science of biology forever. His *theory of natural selection* includes the following premises:

- Organisms produce more offspring than can be supported by the environment.

- Among these organisms there are variations, and these variations may be inherited.

- Some variations allow some organisms to survive longer and produce more offspring.

- There is a "struggle for existence" with "survival of the fittest" among the individuals in a population. This process he called natural selection.

- Natural selection over long periods of time may lead to the accumulation of changes that differentiate one species from another and lead to new species.

Darwin's two important phrases have often been misinterpreted over the years. The "survival of the fittest" does not refer to being strong or having a long life, but to surviving and reproducing. Fitness, in biological terms, refers to the passing on of one's genes to the next generation. The most fit organism is the one that produces the most offspring. The "struggle for existence" refers not to open combat among organisms, but to the organism's ability to use its environment to survive and reproduce. For example, some birds may be more aggressive in declaring a territory and more agile in obtaining nesting materials and food supplies; these birds will survive and produce more offspring.

Darwin expanded his premises to state that all forms of life developed gradually and by chance, over long periods of time, from different and often simpler ancestors; indeed, he stated that all branches of life lead back to a common ancestor. An evolutionary tree of life is a diagram of this principle of common ancestry.

Darwin was not the first scientist to point out the demonstrable fact that living things, and the populations that they are part of, gradually change. Darwin's legacy is that he proposed a mechanism for *how* that change occurs—the theory of natural selection. It is important to note that the common usage of the term "theory" in everyday life, as a speculative idea or mental viewpoint, is not the way the word is used in science. Instead, scientists use the term theory for an important idea that is backed by considerable supporting evidence.

A MODERN INTERPRETATION

Since Mendel's paper on the inheritance of traits was not discovered until 1900, Darwin had no knowledge of genes or genetic principles, such as why parental traits are not blended in the next generation, or why traits can disappear in one generation and reappear in the next. Research conducted in the twentieth century has offered insights into the factors that influence natural selection. As you will remember from your reading of "The Wonderful Mistake" in Learning Experience 3— The Language of Heredity, it is "errors" in DNA that drive evolution. How do DNA mutations help explain the factors that underlie Darwin's premises, premises that led him to develop the theory of natural selection and change the way we look at the natural world?

▶ ANALYSIS

Write responses to the following in your notebook.

1. Why do most organisms produce more offspring than can survive?

2. What role does the environment play in the process of natural selection?

3. Name and briefly explain at least three biological factors that can cause variation among offspring and in future generations.

4. How are the finches of the Galapagos an example of natural selection? What was the selection pressure?

5. Evolution, by definition, never ceases. How might you explain the emergence of new microbes, such as the hantavirus and the Ebola virus, according to the theory of natural selection?

6. Evolution may occur also relatively rapidly, particularly in tiny organisms that reproduce rapidly and in huge numbers. It has been said, "Try not to get sick in a hospital—it can kill you!" How might you explain bacterial resistance to antibiotics?

TEDDY GRAHAM SELECTION

Adapted from "Natural Selection with Teddy Grahams" by Robert R. Blake, Jr., Albert C. Wartski, and Lynn Marie Wartski. American Biology Teacher, *March 1993, pp. 64–65.*

INTRODUCTION When you look at the natural world, you see easily that organisms are well adapted to their environments. Many have colors and textures that camouflage them from predators, others have unique structures that enable them to feed efficiently, sense danger or flee quickly. Trees in colder climates have narrow leaves that tend to remain year round, while those in warmer climates have broader leaves that drop in the autumn. There are many such patterns in the living world. However, an organism can not change in and of itself to become better adapted. All such change, such as an animal being born albino (with white skin and fur) occurs at first by chance. The change is acted upon by the environment; albinism may have a selective advantage in the Arctic, but not along a river bank in the tropics. Thus, those organisms having the advantageous variant of a trait will be selected for and reproduce. Organisms may vary, but it is a population that evolves.

In this activity, you will be the environmental factor that applies selective pressure on the organism known as the Teddy Graham. There are two types of Teddy Grahams, the plain and the chocolate. You are a bear-eating predator who prefers to eat only chocolate bears. (It just so happens that the chocolate bears taste good and are easier to catch than the plain bears which taste sour and are harder to catch.)

▶ MATERIALS NEEDED

For each pair of students:
- 10 Teddy Grahams in a plastic bag (plain and chocolate)
- 1 sheet of graph paper
- paper towels

For the class:
- additional Teddy Grahams

▶ PROCEDURE

1. Obtain a mixed population of Teddy Graham bears from your teacher.
2. Copy Table 10.2, located at the top of the next page, into your notebook.

GENERATION	CHOCOLATE BEARS	PLAIN BEARS	TOTAL BEARS
1			
2			
3			
4			

Table 10.2

3. Place your bears on a paper towel and record the numbers of chocolate bears, plain bears, and the total population. This is generation one.

4. Eat three chocolate bears. (If you do not have three chocolate bears, then eat the difference in plain bears.)

5. Each surviving chocolate bear produces one new chocolate bear, and each plain bear produces one new plain bear. Obtain the new generation of bears from your teacher, and record these numbers as the start of generation two.

6. Repeat Steps 4 and 5 for two more generations.

7. Create a bar graph for the totals of each type of bear in each generation.

▶ ANALYSIS

Write your response to the following in your notebook.

1. Use your knowledge of natural selection to explain what happened to each population of bears over the four generations.

2. What do you think will happen to each type of bear over 10 or 20 generations? Why?

POPULATION GENETICS

READING

Have you wondered why recessive alleles do not disappear from a population, even when they can lead to fatal disorders? Or why some genes seem more frequent than others in a certain population? The field of biology referred to as *population genetics* is a synthesis of the principles of evolution and genetics. A population is a group of organisms of the same species living in a given area. (A species is a group of organisms that can interbreed and produce fertile offspring, for example, dogs.) To a geneticist, a population is defined by its *gene pool,* the sum total of all the alleles of all the individuals in that population. Geographic barriers tend to keep a population in a given area. As the individuals there inter-breed and new generations are produced, some alleles will be selected for and increase while other alleles may be selected against and decrease. *Thus, evolution can be defined in a new way as the change in the gene pool of a population over time.*

How often a particular allele appears in a gene pool is referred to as the *gene frequency*. Frequencies may vary within the same species, depending on the location of the specific population. For example, the fish in a lake in Massachusetts may have a different proportion of certain alleles than the same species of fish in a lake in Missouri.

TAY-SACHS DISEASE

What about the human gene pool? Subgroups or races evolved hundreds of thousands of years ago and have their own gene pools. This has changed somewhat with the accessibility of travel and some interbreeding among races and ethnic groups. However, the gene pools of subgroups remain relatively constant. The evidence of unusual allele frequency within a subgroup is of great interest to population geneticists. For example, Tay-Sachs disease results from inheriting two recessive alleles that cause a fatal degeneration of the nervous system, usually by age four. It is found relatively often among Ashkenazi Jews (Jews of Central and Eastern European origin). The frequency of the allele is as follows:

- In the general U.S. population, 4 persons in 1,000 carry the allele.
- Among Ashkenazi Jews worldwide, 28 in 1,000 carry the allele.
- Among New York City Jews, 33 in 1,000 carry the allele.

SICKLE CELL AND MALARIA

Mutations, as you know, are one of the causes of genetic variability. If the resulting phenotype is advantageous, it will be selected for; if it is disadvantageous, it will be selected against and the organisms carrying the mutation usually will be weakened or die. However, sickle cell is a notable example of a harmful mutation having a selective advantage in regions plagued by malaria. Why is this?

The first systematic study of this question was carried out in 1954 when Tony Allison, a scientist working in England and Africa, examined the distribution of the sickle cell trait. He observed that the trait was mainly confined to tropical Africa where up to 45% of the individuals in some tribes had the trait and about 3% of the children suffered from sickle cell anemia. The gene was also present in the Negroid Veddoid aboriginals of India, the Achdam of southern Arabia, and to a lesser extent, in some Mediterranean peoples. Remembering some earlier information he had seen about the distribution of a parasitic disease, malaria, which is caused by the protozoan *Plasmodium falciparum*, Allison made a connection that resulted in the map shown in Figure 10.3.

Malaria is a disease carried by the *Anopheles* mosquito and spread by the bite of the mosquito which injects a parasite into the bloodstream of the human host. The *Plasmodium* parasite infects red blood cells causing them to burst every three days. The major symptoms of malaria are the alternating fevers and chills as the red blood cells burst and leave the victim weak with headaches, nausea, anemia, and a loss of appetite. In areas of the world where the *Anopheles* mosquito is present, malaria is a serious disease—250 million people suffer from it yearly and more than two million die.

As you saw in Denzel's family in Learning Experience 2—No Matter What Your Shape, sickle cell comes in two forms; the heterozygote is called sickle cell trait and the recessive is sickle cell anemia. It is thought that the gene that codes for the beta-hemoglobin chain mutated in Africa thousands of years ago, and originally was probably very rare among the population. But, as farmers cleared the land for crops, breeding sites for the *Anopheles* mosquito were probably inadvertently produced, and at the same time the gene frequency of the sickle cell allele rose sharply. Considering that untreated victims of sickle cell anemia often died, why was this harmful allele increasing? Why wasn't the allele being selected against?

Researchers discovered that heterozygotes for sickle cell are less susceptible to malaria than homozygous dominants (who do not carry the sickle cell allele and are of normal genotype). Individuals who are homozygous dominant are, in fact, less fit for their environment in that they are more likely to contract malaria and die. Since heterozygotic individuals are more likely to survive and produce offspring, the recessive allele increased in the populations where malaria was present. Thus, having the allele was advantageous and, since it was a deterrent to a deadly disease, it was selected for throughout the equatorial regions of the world (see Figure 10.3).

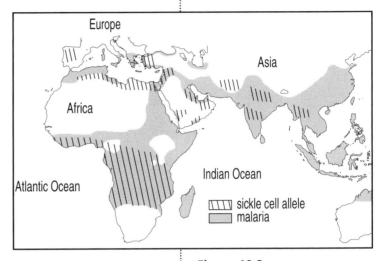

Figure 10.3
Geographic distribution of malaria and sickle cell.

▶ ANALYSIS

Write responses to the following in your notebook.

1. Explain the frequency of the Tay-Sachs allele in the groups presented.

2. If the sickle cell allele offers no advantage to an African-American carrier in the United States where malaria is not a problem, do you think it will be selected against and disappear in this country? Explain your response.

WHAT'S IN OUR GENE POOL?

INTRODUCTION As you have seen in the last learning experience, human traits generally result from the interaction of several genes, and the mode of inheritance is often more complicated than originally thought. However, some traits are easily seen and explained in genetic terms. In this activity, you and your partner will determine your phenotypes for several selected traits. You will record your data and form a table which will include the percentage of class members having each phenotype.

▶ TASK

1. Carry out steps 2–4 and list each of your phenotypes in your notebook.

2. Attempt to roll your tongue by curling the sides of your tongue up. Record whether you can or cannot.

3. Attempt to fold your tongue over. Record whether you can or cannot.

4. Note and record the presence or absence of the following inherited traits:
   a. Do you have free or attached ear lobes (see Figure 10.4a)?
   b. Do you have a widow's peak or a straight hairline (see Figure 10.4b)?

Figure 10.4

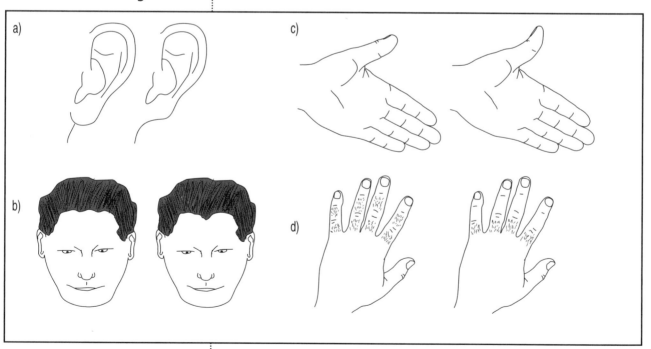

c. Do you have a straight thumb or a hitchhiker's thumb (see Figure 10.4c)?

d. Do you have hair on the middle portion of each finger or no hair (see Figure 10.4d)?

e. Do you have 5 or 6 fingers on each hand?

5. Enter your phenotypes in the class data table on the board.

6. Use Table 10.5 as a guide to record your genotype next to your phenotype.

Table 10.5

▶ HUMAN TRAITS	DOMINANT	RECESSIVE
tongue rolling	can	cannot
tongue folding	can	cannot
shape of ear lobe	free	attached
hairline	widow's peak	straight
shape of thumb	straight	hitchhiker's
hair on mid-digit	present	absent
number of fingers on each hand	six	five

7. Record the percentage of the class having each dominant trait.

▶ ANALYSIS

Write responses to the following in your notebook.

1. What surprised you about Table 10.5? Why?

2. What surprised you about the percentages of students having the dominant trait? What do you think the explanation might be?

3. How might you explain some of the data in Table 10.6, using the principles of natural selection?

Figure 10.6
Table showing dominant and recessive human traits.

▶ RECESSIVE	DOMINANT
type O blood	type A or B blood
normal hip joints	dislocated hip birth defect
blue eyes	brown eyes
normal eyelids	drooping eyelids
no tumor of the retina	tumor of the retina
normal fingers	short fingers
normal thumb	extra joint in the thumb
normal fingers	webbed fingers
ability to smell	inability to smell
normal number of teeth	extra teeth
presence of molars	absence of molars
normal palate	cleft palate

Natural Detours

Natural Selection, by Ensuring that Advantageous Traits Are Passed on to the Next Generation, Drives Evolution. Why Is It Then That Genes So Often Get in the Way?

Carl Zimmer, ©1996 The Walt Disney Co. Reprinted with permission of Discover Magazine [October 1996, page 24].

Is natural selection the prime mover behind evolution? Darwin's great insight into the mechanics of evolution was that a population of creatures always has a lot of variation—more feathers here, less fat there, more urge to kill there—and some of these variations allow the individuals bearing them to thrive and have more offspring than others. After many generations these

mine fitness, biologists have always suspected that natural selection seldom guides a species straight up the evolutionary heights. It has been hard, though, to find data to support such suspicions—until now.

Biologist Dolph Schluter of the University of British Columbia has been studying three-spined sticklebacks, a genus of fish that lives in western Canada. When glaciers pulled back from the region 13,000 years ago, sticklebacks invaded the newly carved lakes and over time formed new species. Schluter captured some individuals from one species that has evolved into a big, bulky, and mean-looking form that makes its living by sucking sediment into its wide mouth. Schluter measured a number of traits important to its way of life and then bred the fish. When their offspring matured, he repeated the measurements. Naturally, the mixing of genes meant that the second generation was not a carbon copy of its parents—the range of different forms was an

expression of the built-in variability of stickleback genes.

But the variations between the generations were not random; instead, all the traits varied together. A stickleback that was unusually long was also unusually fat and had a wide mouth. Similarly, a short stickleback was invariably slender and had a narrow mouth. Variation works like this because of the way genes build our bodies. A wide mouth is the product of many genes, not just one, and many of these genes play a part in creating other traits as well.

The odd thing about this pattern of variability among individual sticklebacks is that the species as a whole has evolved in the same manner. The fat, wide-mouthed species that Schluter studied descended from a shorter, slimmer, smaller-mouthed species that first invaded the region's lakes. The traits of shortness, slimness, and narrow mouths have remained linked in the sticklebacks for at least 13,000 years. Even though natural selection might favor, say, a long, slender, wide-mouthed stickleback, the linked genes for these traits prevent—for millennia—such a form from arising.

The connection between

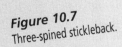

Figure 10.7
Three-spined stickleback.

traits become more common among the population as a whole. If one imagines fitness as a beckoning peak on a given ecological landscape, then natural selection should be a process that moves species steadily uphill.

Countless experiments have proved that natural selection is real and quite powerful. Yet given the complexity of the interplay of genes that deter-

evolution and variability was no coincidence, Schluter found. He looked at similar measurements taken from other species, such as sparrows, finches, and mice. In every case, evolution was biased toward the kinds of body shapes produced most easily by the variability of the genes. For some of the species he studied, the genetic constraints persisted for 4 million years. This genetic rein on natural selection, Schluter's work shows, is far more persistent than anyone had thought.

Schluter likes to call the bias "the genetic path of least resistance." As natural selection tries to change a feature on an animal—say, the mouth's width—other traits change with it because the same genes control many traits. "It's easier to change along some directions than others," says Schluter.

Over very long periods of time, says Schluter, natural selection does eventually break down the constraints imposed by such linked genetic traits. It tries to steer the species steadily upward, but linked variability inevitably delays this upwardly mobile tendency for millions of years. "We do expect a species to get to the point where selection is directing it," says Schluter. "It's just going to go in a roundabout way."

▶ ANALYSIS

Write a short essay on your interpretation of the title of this learning experience—Genetics and Evolution: The Needle and the Thread.

EXTENDING IDEAS

▷ In 1989, the first human gene, that for the disease cystic fibrosis, was sequenced; it is on chromosome 7. This debilitating disease, whose symptoms include salty sweat and thick mucus that clogs air passages and the digestive gut, is the most common fatal genetic disorder among Caucasians; 1 in 2000 have the disease and 5% of the white population carry one copy of the mutant gene. Even today, with antibiotics and other treatments, those afflicted rarely live past forty.

In 1994, geneticists from Barcelona, focused on a set of markers in the gene sequence and determined that the gene was the result of a mutation that occurred in East Asia at least 52,000 years ago, before the migration of peoples northward into Europe and into other areas. Why has this mutant recessive gene survived when it appears obvious that it should have been selected against long ago? Could it offer a selective advantage for something? Why did the cystic fibrosis mutant in all its forms survive only among white European and their descendants who immigrated to the United States?

It was reported, also in 1994 from experiments with mice, that the cystic fibrosis mutation offers increased resistance to the diarrhea that results from cholera, a bacterial disease that has killed millions of people over past centuries. This protective effect was also thought to extends to other diarrheal diseases caused by bacteria. Yet, white populations that had emigrated to areas such as South America seemed not to have the mutation and large numbers of them had dies in cholera epidemics. Why didn't it survive in those warmer regions? It was speculated that in hot climates there was an additional disadvantage to having the gene, one that was more important than diarrhea—salty sweat. Experiments have shown that carriers of one mutant gene lose more salt in their sweat than than those with two mutant genes. In a human who hunts or travels with cattle, losing salt was serious because it was not easily replaced in the body.

The cystic fibrosis gene is another instance of selective pressure affecting frequency. You may wish to explore further the following topics:

- how the mutant gene affects the cells of a person with cystic fibrosis
- how the action of the gene protects against diarrhea in both those with the disease and in carriers
- the methods used to date the mutation of the gene
- the mechanisms of natural selection in preserving and eliminating the gene in different areas

▶ There are three main patterns of evolution—divergent, convergent, and parallel. Research one of these patterns and give examples of species that follow that pattern.

▶ Find out about a particular species that has been listed as threatened or endangered and whose gene pool is a factor in its status. What is the range in its gene pool? Can the gene pool be expanded, and if so, how? What do you think will happen to this species over time?

▶ Conduct research on one of the species Darwin studied. How did he think the evolution of that species occurred?

▮ON THE JOB

WILDLIFE MANAGER   Growing up, Lisa always had a love of animals. She thought she wanted to be a veterinarian and work with farm animals. But in high school, she heard more and more about many species—the panda, the bald eagle, the California condor, the right whale—being in danger of extinction. Hearing all the news accounts about endangered species made her want to do something more than just send a donation. To learn more about the problem, she took biology courses and any environmental courses that were offered at her high school. When Lisa went to college, she looked for a school with a strong biology program. Again, she added classes in environmental sciences. Then she enrolled in a master's program in wildlife management.

Lisa is now a wildlife manager working for an international organization whose primary concern is to conserve wildlife species. One of her largest projects is the revival of the green sea turtle species. Sea turtle hatchlings are sometimes lured away from the ocean towards the bright lights of cities and towns bordering the sea. They become quick prey for birds and other predators, or they starve to death without the nutrients they would receive in their ocean habitat. Lisa has been working with local government officials in different coastal communities, developing a proposal that will protect the baby turtles and encourage them to use their natural instincts to head toward the ocean while not hindering greatly the residential or commercial progress of these areas. Lisa is also looking out for the welfare of adult sea turtles which are regularly trawled up in commercial shrimping nets. She has been working for legislation toward only allowing "sea turtle safe" shrimp to be imported into the United States. This practice, alone, could save thousands of turtles each year.

Although extinction can be thought of as a part of the evolutionary cycle, because of human interference in many animal habitats the rate of extinction is skyrocketing. But more and more people like Lisa are becoming aware of the problem and trying to make a difference.

Wildlife managers have to set their sights on the far future. Sometimes, increasing the populations of endangered species can take years—even more than a century. In order to successfully manage wildlife, one must take into account more than just whether the animal is endangered. One must look at all aspects of these animals' lives, including the habitat in which the species live (Is it being polluted by pesticides or man-made waste? Has a non-native predator been introduced to the area? Are the animals coming in contact with man-made obstructions such as electrical wires? Is the habitat disappearing because of industry?), its source of food (Is its food supply polluted or disappearing?), and its potential commercial uses (Are humans fishing it into extinction?).

As numbers of an animal species diminish, the chances of inbreeding increase. Inbreeding also increases the chances of genetic diseases, thus weakening the gene pool. In order to strengthen the remaining specimens, wildlife managers may work with other scientists and try to breed endangered species; for instance, Lisa may work with zoos to bring pandas together for mating in captivity.

Some organizations have taken even more drastic steps to keep species around for the future. Although the technology hasn't been developed yet, at least one zoo in the western United States is freezing species' DNA in the hope that if these species become extinct, there will be some way to recreate the disappearing animals from their genetic material.

Lisa views her work as not just a job, but a way of life. In high school, she had doubts that she could make a difference, but she took her belief in the need to stop the rate of animal extinction and has made it into a cherished accomplishment. Although changes may not come easily or quickly, she feels that saving even a few animals now can make a big difference in the long run.

A CURRENT AFFAIR

PROLOGUE **A**s you have read genetics headlines and articles in this module, have you thought about what each might mean to you now and in the future? In this learning experience, you will look closely at an article on a topic in genetics of your choosing. What does the article reveal about the science upon which the article is based? Often very important scientific topics are paraphrased and presented in ways that make them easier for the public to understand. What information might you need to understand the science more completely? What does the article say about the issues and concerns different groups might have regarding the topic? How might the technology or the discovery mentioned in the article affect you now or in the future?

With your article as a starting point, you will investigate this topic in depth, describe its relationship to your life, and write an independent research paper that includes a case study.

CONSIDER THE SOURCE

INTRODUCTION   Newspapers, magazines, television, radio, and now the World Wide Web are our major sources of information about emerging developments in scientific and technological fields. What do *you* think about when you read or hear these reports? What makes a story interesting and relevant to you? Has studying genetics made you more aware of the number of discoveries and technological advances being reported almost daily?

You will need to write a paper that looks closely at what is currently being reported in a particular field of genetics research and technology and analyze the impact this new information may have on *your* life. The following Task is divided into three major components: You will be responsible for choosing an article on genetics that is important to you, doing research on the topic in the article, and writing a report which includes scientific background, a case study, and a personal statement. This is an independent study, requiring a lot of organization, creativity, and thoughtfulness on your part. As you read through the Task, your teacher will talk with you about the report and the amount of time you will have to complete it. You will be evaluated on the report.

▶ TASK

1. Search newspapers, magazines, and if you have access, the World Wide Web for an article on a genetics issue that you feel relates to your life either now or in the future. The article can deal with a genetic disease, genetics research, genetic technology, or environmental influences that can alter gene expressions.

   a. Be creative when thinking about how the topic in the article might relate to you now or in the future, whether the relationship is direct or through effects on your family or society. For example, if you are a vegetarian you may be interested in the issue of whether a vegetarian would eat or drink products with animal genes in them. Or, your grandfather may have a genetic disorder, and so you may want to find about what is known about the disorder and what it may mean for you. You may live on a farm or ranch where gene technology to enhance certain traits in plants or animals is important, and you may find an article about the genetics of those traits. These are just a few possibilities.

   b. As you read newspapers and magazines looking for a topic that would be interesting and relevant to you, focus your thinking on determining why it is interesting to you. Once you choose an article, write in your notebook questions that the article raises, issues and concerns that might relate to the article, ways in which the article has meaning for you in your life, and the ways the topic might affect your personal values.

2. Find more resources on the topic. Look for information on the science underlying the topic, on any research or technology that has been critical in uncovering information on the topic, and on ethical issues or concerns related to the topic. Use the concepts in this module as a resource for scientific information.

3. Write a research paper that includes the following:

   a. A report that summarizes your research. This synthesis of your research should include deeper scientific knowledge on the topic than was in the original article and, if applicable, a description of the technologies associated with this advancement in genetics.

b. A case study that brings forth the issues involved in this topic. The case study can be written as an article which describes the issues and concerns that relate to the topic and includes your own decision-making based on your analysis of the risks and benefits, *or* it can be an original narrative that describes an individual in a situation and the *issues and concerns* that confront that individual, and what decision you would choose.

NOTE: Look at the case studies that are in Learning Experiences 2, 4, and 7 for examples of ways to present a case study.

c. A short essay which describes how the topic you chose relates to your life and, if appropriate, what this might mean for society at large. The essay should also include how the research for this learning experience may have influenced your opinions and concerns related to the topic.

d. A bibliography of the references you used in writing the report.

e. Attach to your paper the following:
   - A copy of the article you chose with the source and date of publication.
   - The notes you recorded as described in step 1 above.

► ASSESSMENT

Your report will be assessed on the following criteria:

• Have you gathered sufficient and appropriate scientific information?

• Have you reached logical conclusions about the issues and concerns around the genetics topic?

• Does your case study reflect insight and understanding about the science and the issues and include risks and benefits and your decision?

• Does your final essay describe how the topic is related to your life and/or has meaning for society?

• Does your final essay include the values you bring to the topic and ways in which your research may have affected your thinking regarding the topic?

• Is your report organized, clear, and complete? Does your report engage and interest the reader in your topic and the issues that surround it?

• Is your bibliography complete and accurate?

• Have you documented other authors' ideas and words?

EXTENDING IDEAS

ON THE JOB

SCIENCE WRITER Geri had a deadline to reach. Her article for *In 2 Science*, a monthly science magazine, was supposed to describe the latest genetically engineered foods and their future repercussions. She had spent many hours interviewing scientists, nutritionists, and government officials so she could represent the different views and issues that resulted from these new edible creations. She felt that her article was scientifically accurate yet accessible to the public. *In 2 Science* was aimed at a high school audience.

Geri had started writing when she was very young. First it was just a personal journal. Then she wrote for her high school newspaper and even sent some articles into her local newspaper. She also loved science. Biology courses were her favorite, but instead of just a laboratory report, she found herself describing her experiments with journalistic vigor. When high school graduation approached, Geri wondered if there was a way to combine her two loves into one career. Should she major in journalism or in biology?

A few weeks into her first college semester, she sat down with her advisor who suggested that she major in biology and minor in English, although acknowledging that there were other approaches to reach a similar goal. Her advisor used the term "niche journalism" where up-and-coming journalists are becoming more specialized in particular fields, and Geri's specialized field would be biological science. In addition to her coursework, she made time to join the staff of the college paper. Her regular column covered the discoveries occurring in the science laboratories on campus as well as certain scientific topics that may, in the near future, affect all people in some aspect of their lives, for instance, cloning or the availability of genetic tests.

During her senior year of college, Geri learned about internships in science journalism at a local newspaper or magazine. She found a position in the science division at the Ledger where she was expected to write articles that would be added to a monthly "Science in the News" insert. She also learned through the internship and further research that there were many places where she could use her skills after college. Not only were there daily newspapers and science magazines, but there were also very specialized science and medical journals, television and radio stations, trade papers, and even popular magazines which reserved sections for reporting the latest advancements in science. Other options included working in public relations for medical firms where she could write press releases for new products and discoveries. Her internship created many opportunities for Geri to meet well-known writers and make numerous contacts with scientists, technicians, and other people associated with medicine, business, law, etc. Through this job, she met the editor-in-chief of *In 2 Science* who was very impressed with her work. After graduation, she accepted a position on the staff of the magazine.

For her own professional development, Geri hopes to continue with her education and gain a master's degree in journalism. But for now, she will continue learning through experience and writing articles for *In 2 Science.*

GLOSSARY OF TERMS

The following terms can be found on the listed page in the Student Manual unless otherwise noted. ◆ indicates pages which you may receive from your teacher.

adenine	26
allele	68
amino acids	26
anticodon	33
autosomes	98
carriers	18
centromere	52
chromatid	52
chromosomes	50
codominance	106
codon	32
complementary	28
complete dominance	106
crossing over	98
cytosine	26
deoxyribonucleic acid (DNA)	25
deoxyribose	26
dihybrid	90
diploid	62
DNA fingerprinting	102
dominant	◆ 46
evolutionary tree	116
gametes	61
gene frequency	122
gene linkage	95
gene pool	121
gene therapy	100
genes	25
genetic marker	99
genetics	25
genome	98
genotype	65
guanine	26

haploid	62
heterozygous	◆ 46
histones	50
homologous chromosomes	67
homozygous	◆ 46
Human Genome Project	100
incomplete dominance	106
karyotype	53
meiosis	62
messenger ribonucleic acid (mRNA)	31
mitochondrial DNA	◆ 144
monohybrid	89
multiple alleles	106
mutations	41
natural selection, theory	118
nitrogenous bases	26
nucleotide	28
oogenesis	62
pedigrees	20
phenotype	65
phosphate	26
pleiotropic	106
polygenic	106
polypeptide	33
population genetics	121
principle of dominance	87
principle of independent assortment	87
principle of segregation	87
Punnett square	88
purine	26
pyrimidine	26
recessive	86
recessive	◆ 46
ribonucleic acid (RNA)	31
ribose	31
ribosome	33
sex-linked genes	99
sickle cell anemia	17
spermatogenesis	63
thymine	26
trait	1
transcription	33
transfer RNA (tRNA)	33
translation	33
translocation	55
trisomy	55
uracil	31
variants	13